# PREFACE OF THE FIRST EDITION

Now a day in fast track jobs it's very important to prepare the detailed and realistic program to achieve the project completion date. As we are aware that MEP activities are very important for successful timely completion of project. We have experienced in our practical life that most of the projects are delayed because of delay in material delivery, installation, testing & commissioning of major MEP activities.

This Mechanical, Electrical and Plumbing (MEP) Planning Manual, has been prepared to highlight all MEP related detailed activities for a typical multistoried tower and villa. In this edition the major concern is to identify the major fields and sequence of activities required to be implemented in order to render the completion of various services to successful operation.

This manual is divided into two parts i.e.

| Part-I | : | MEP Activity Details |
| Part-II | : | MEP Method Statement |

The main purpose of this MEP Planning Manual is to prepare a program of works with detailed activities for all MEP works integrated with thin the main civil program.

This manual has been prepared by HEE MEP Planning Engineer Mr. Gulshan Kumar, reviewed by HEE Planning Manager Mr. Jafar M. Khair.

This manual will be review periodically as required for adjustment and enhancement.

# Table of Contents
## Part-I
## Activity Details

| | |
|---|---|
| PLANNING DEPARTMENT | |
| Rev.: 0 | **MEP PLANNING MANUAL** |
| Date of Issue: Feb 2007 | |
| Doc No.: MEP-01-07 | |

**www.arabmep.com**

3

# *Table of Contents*
# *Part-II*
# *Method Statements*

| PLANNING DEPARTMENT | | |
|---|---|---|
| Rev.: 0 | **MEP PLANNING MANUAL** |  |
| Date of Issue: Feb 2007 | | |
| Doc No.: MEP-01-07 | | |

**www.arabmep.com**

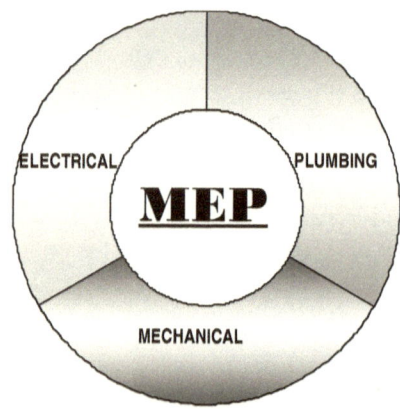

# MEP PLANNING MANUAL

## (PART – I)

# *A Guide to the Project Planning & Installation related to MEP Works*

### *First Edition (Feb, 2007)*

# SCOPE OF MEP WORKS

The MEP (Mechanical, Electrical & Plumbing) works is mainly divided into the following main categories:-

- **Mechanical:-**

  The Mechanical installations mainly cover HVAC works.

- **Electrical:-**

  The Electrical Installations is covering all electrical works, fire alarm system, telephone system, BMS, low voltage system.

- **Plumbing:-**

  The Plumbing Installations is covering drainage system, water supply distributions, sanitary ware installations and related works.

- **Fire Fighting System:-**

  The fire fighting system is covering sprinkler system, fire fighting equipments, clean agent FM200 system, etc.

# SECTION 1

## MEP ACTIVITY LIST

This covers all the MEP activities in detail, describing 1st 2nd & final fix. Each activity has been given an individual code number.

This code number can be used as a basic reference code while preparing the MEP Program. The first two digit of activity code is representing the system:-

     KY       :   Project Milestones

     PL       :   Plumbing System

     FF       :   Fire Fighting System

     AC       :   HVAC System

     EL       :   Electrical System

Each service has separate activity lists i.e. plumbing, fire fighting, electrical, HVAC.

In the last column you will find the relevant method statement reference no. for the particular activity. Refer these method statements to know more detail about the activity installation procedures, quality of work, safety requirement implementation and resources (material, equipment & personnel).

# PROJECT MILESTONES REQUIREMENTS

| Type | Area | Code | Description |
|------|------|------|-------------|
| Milestones | Tower | KY-01 | Temporary Power Availability |
| | | KY-02 | Drainage Connection |
| | | KY-03 | Water Availability |
| | | KY-04 | Civil Defense Inspection |
| | | KY-05 | Authority Inspection (Electrical) |
| | | KY-06 | Permanent Power Availability |
| | | KY-07 | Chilled Water Availability (from District Cooling Plant) |
| | | | |

# PLUMBING SYSTEM

| Type | Area | Code | Description |
| --- | --- | --- | --- |
| Underground | Typ. Floor | PL-01 | Underground Piping |
| Embedded in Concrete | Typ. Floor | PL-02 | Putting sleeves in concrete slab |
| | | PL-03 | Embedded pipe in concrete slab |
| 1st Fix | Typ. Floor | PL-04 | Cutting & Chasing in block walls |
| | | PL-05 | Putting Pipes (water supply / drainage) in block walls |
| 2nd Fix | Typ. Floor | PL-06 | Horizontal Drainage Piping |
| | | PL-07 | Horizontal Water Supply Piping |
| | | PL-08 | Drainage Pipe Water Leakage Testing |
| | | PL-09 | Water Supply Water Leakage Testing |
| | | PL-10 | Vertical (Riser) Drainage Piping |
| | | PL-11 | Vertical (Riser) Water Supply Piping |
| | | PL-12 | Floor Drain (Trap) Fixing |
| | | PL-13 | Water Heater Installation |
| | | PL-14 | Bath Tub / Shower Tray Installation |
| | Basement / Mech. Floor / Roof | PL-15 | Sewage Lift Pump Installation & Piping |
| | | PL-16 | Water Transfer Pump Installation & Piping |
| | | PL-17 | Water Booster Pump Installation & Piping |
| | | PL-18 | Erection of Water Tank & Connections |
| Final Fix | Typ. Floor | PL-19 | Sanitary Ware Installation |
| | | PL-20 | Sanitary Fixture Fixing |
| | | PL-21 | Water Meter Fixing |
| Testing & Pre-Commissioning | Tower | PL-22 | Pre commission water transfer & Booster pumps |
| | | PL-23 | Pre commission floor water heaters |
| | | PL-24 | Overhead Roof Water Tank fillup |
| | | PL-25 | Flushing & checking leaks in WS pipes |
| | | PL-26 | Flushing & checking leaks in drainage pipes |
| | | PL-27 | Simultaneous flow test on drainage risers |
| Commissioning & Handover | Tower | PL-28 | Test and commission water transfer & booster pumps |
| | | PL-29 | Commissioning water supply PRV's |
| | | PL-30 | Commissioning floor water heaters |
| | | PL-31 | Commissioning of bath tubs & sanitary ware |

# FIRE FIGHTING SYSTEM

| Type | Area | Code | Description |
|---|---|---|---|
| 1st Fix | Typ. Floor | FF-01 | Fire Fighting Hangers & Supports |
| 2nd Fix | Typ. Floor | FF-02 | Fire Fighting Piping |
| | | FF-03 | Fire Fighting Pipe Riser |
| | | FF-04 | Fire Fighting Pipe Testing |
| | | FF-05 | Fire Fighting Accessories Fixing |
| | Basement / Mech. Floor | FF-06 | Fire Pump Set Installation |
| | | FF-07 | Fire Pump Set Pipe Connections |
| Final Fix | Typ. Floor | FF-08 | Sprinkler Fixing |
| | | FF-09 | Fire Hose Cabinet Fixing |
| | | FF-10 | Fire Hose Reel Fixing |
| | | FF-11 | Fire Extinguisher Fixing |
| | | FF-12 | Fire Blanket Fixing |
| Testing & Pre-Commissioning | Tower | FF-13 | Pressurize Fire Fighting pipes |
| | | FF-14 | Flushing & Refilling of Fire Fighting Pipes |
| Commissioning & Handover | Tower | FF-15 | Commissioning Fire Fighting PRV's |
| | | FF-16 | Commissioning of FHC's |
| | | FF-17 | Commissioning of Sprinklers |

# AIR CONDITIONING SYSTEM (HVAC)

| Type | Area | Code | Description |
|---|---|---|---|
| 1st Fix | Typ. Floor | AC-01 | Duct hangers & Supports |
| | | AC-02 | Chilled Water Pipe Hangers & Supports |
| 2nd Fix | Typ. Floor | AC-03 | Ducting Supply, return, Fresh & Exhaust |
| | | AC-04 | Volume, Fire & Smoke Damper Fixing |
| | | AC-05 | Ducr Risers |
| | | AC-06 | Duct Insulation |
| | | AC-07 | Chilled Water Piping |
| | | AC-08 | Chilled Water Pipe Riser |
| | | AC-09 | Chilled Water Pipe Test |
| | | AC-10 | Chilled Water Pipe Insulation |
| | | AC-11 | FCU Installation |
| | | AC-12 | Duct Connection to FCU |
| | | AC-13 | Pipe Connection to FCU |
| | | AC-14 | Bends, Droppers & Mouth Piece |
| | Basement / Mech. Floor | AC-15 | Chillers Installation |
| | | AC-16 | Heat Exchanger Installation |
| | | AC-17 | Staircase Pressure Fan Installation |
| | | AC-18 | Garbage Exhaust Fan Installation |
| | | AC-19 | Chilled Water Pump Installation |
| | | AC-20 | Expansion Tank & Pressure Unit Installation |
| | | AC-21 | Air Scrubbers installation |
| | | AC-22 | Chemical Dozing Plant Installation |
| | | AC-23 | AHU Installation |
| | | AC-24 | Duct Connection to AHU |
| | | AC-25 | Pipe Connection to AHU |
| Final Fix | Typ. Floor | AC-26 | Kitchen Exhaust Fan Fixing |
| | | AC-27 | Ceiling Diffuser Fixing |
| | | AC-28 | Grille & Linear Fixing |
| | | AC-29 | Thermostat Fixing |
| Testing & Pre-Commissioning | Tower | AC-30 | Pre commissioning chilled water pumps |
| | | AC-31 | Water circulation in system and drain (Static) |
| | | AC-32 | Circulation of water with cleaning chemical |
| | | AC-33 | Pre commissioning FAHU's |
| | | AC-34 | Cleaning of all filters and strainers |
| | | AC-35 | Pre commissioning pressurization fans |
| | | AC-36 | Water circulation in system and drain (dynamic) |
| | | AC-37 | Refilling of water and witness by Engineer |
| | | AC-38 | Prelim. Adjusting and balancing of valves |

# AIR CONDITIONING SYSTEM (HVAC Cont…..)

| Type | Area | Code | Description |
|---|---|---|---|
| | | AC-39 | Test and commission chilled water pumps |
| | | AC-40 | Test and commission heat exchangers |
| | | AC-41 | Test and commission FAHU's |
| | | AC-42 | Test and commission floor FCU's |
| Commissioning & Handover | Tower | AC-43 | FAHU air balancing |
| | | AC-44 | Air balancing supply & return grilles |
| | | AC-45 | FCU water balancing |
| | | AC-46 | Balancing air from pressurization fans |
| | | AC-47 | Water balancing of FAHU's |
| | | AC-48 | Main System Valves adjustment |

## ELECTRICAL SYSTEM

| Type | Area | Code | Description |
| --- | --- | --- | --- |
| Embed. in Conc. | Typ. Floor | EL-01 | Embedded Conduits in Concrete |
| 1st Fix | Typ. Floor | EL-02 | Marking on Block walls |
| | | EL-03 | Cutting & Chasing in block walls |
| | | EL-04 | Fix Conduits, GI Boxes in Block Walls |
| 2nd Fix | Typ. Floor | EL-05 | G.I. & PVS Surface Conduiting |
| | | EL-06 | Installation of Distribution Boards |
| | | EL-07 | Installation of Sub-Main Boards |
| | | EL-08 | Installation of Main LV Panel |
| | | EL-09 | Erect Cable Trays |
| | | EL-10 | Cable Trucking |
| | | EL-11 | Earthing System |
| | | EL-12 | LV Cabling |
| | | EL-13 | ELV Cabling |
| | | EL-14 | Main LV Cable Termination |
| | | EL-15 | Sub main Cable Termination |
| | | EL-16 | Wiring for Lighting & Small Power |
| | | EL-17 | Wiring for ELV System |
| | | EL-18 | Wiring for Fire Fighting System |
| | | EL-19 | Circuit Continuity Test |
| | | EL-20 | Ceiling Rose & Flexible Drops |
| | | EL-21 | Distribution Boards Dressing |
| | | EL-22 | Bus bar Installation |
| | | EL-23 | Lightning Protection System |
| | Basement | EL-24 | Transformer Installation |
| | | EL-25 | Generator Installation |
| | | EL-26 | ATS Panel Installation |
| | | EL-27 | LV Cabling from Transformer to LV Panel |
| Final Fix | Typ. Floor | EL-28 | Fixing Fire Alarm & Light Fittings |
| | | EL-29 | Fixing Switches, Sockets & Accessories |
| Testing & Pre-Commisssioning | Tower | EL-30 | Circuit continuity test by bus riser |
| | | EL-31 | Continuity test LV wiring |
| | | EL-32 | Pre commissioning ELV fixtures |
| | | EL-33 | Final test and commissioning bus risers |
| Commissioning & Handover | Tower | EL-34 | Test and commission variable frequency drive |
| | | EL-35 | Lighting protection earth resistance test |
| | | EL-36 | Floor MDB, SMDB, MCC panel commissioning |
| | | EL-37 | Test and commission light fixtures |
| | | EL-38 | Test and commission ELV final fixtures |

# MECHANICAL FLOOR / ROOF ACTIVITIES

| Sr. No. | Activity Description |
|---|---|
| 1 | Chillers Installation |
| 2 | Heat Exchanger Installation |
| 3 | Staircase Pressure Fan Installation |
| 4 | Garbage Exhaust Fan Installation |
| 5 | Chilled Water Pump Installation |
| 6 | Expansion Tank & Pressure Unit Installation |
| 7 | Air Scrubbers installation |
| 8 | Chemical Dozing Plant Installation |
| 9 | AHU Installation |
| 10 | FCU Installation |
| 11 | MCC Installation |
| 12 | Booster Pump Installation |
| 13 | Water Tank Installation |
|  |  |
|  |  |
|  |  |

# LIST OF TESTING AND COMMISSIONING ACTIVITIES
## ELECTRICAL

| Sr. No. | Activity Description |
|---------|----------------------|
| 1 | Circuit continuity test by bus riser |
| 2 | Continuity test LV wiring |
| 3 | Insulation resistance test (cabling) |
| 4 | Pre commissioning ELV fixtures |
| 5 | Final test and commissioning bus risers |
| 6 | Test and commission variable frequency drive |
| 7 | Lighting protection earth resistance test |
| 8 | Floor MDB, SMDB, MCC panel commissioning |
| 9 | Test and commission light fixtures |
| 10 | Test and commission ELV final fixtures |

# LIST OF TESTING AND COMMISSIONING ACTIVITIES
# PLUMBING

| Sr. No. | Activity Description |
|---|---|
| 1 | Pre commission water transfer & Booster pumps |
| 2 | Pre commission floor water heaters |
| 3 | Overhead Roof Water Tank fillip |
| 4 | Flushing & checking leaks in WS pipes |
| 5 | Flushing & checking leaks in drainage pipes |
| 6 | Simultaneous flow test on drainage risers |
| 7 | Test and commission water transfer & booster pumps |
| 8 | Commissioning water supply PRV's |
| 9 | Commissioning floor water heaters |
| 10 | Commissioning of bath tubs & sanitary ware |

# LIST OF TESTING AND COMMISSIONING ACTIVITIES
# FIRE FIGHTING

| Sr. No. | Activity Description |
|---|---|
| 1 | Pressurize Fire Fighting pipes |
| 2 | Flushing & Refilling of Fire Fighting Pipes |
| 3 | Commissioning Fire Fighting PRV's |
| 4 | Commissioning of FHC's |
| 5 | Commissioning of Sprinklers |

# LIST OF TESTING AND COMMISSIONING ACTIVITIES
## HVAC

| Sr. No. | Activity Description |
|:---:|:---|
| 1 | Pre commissioning FAHU's |
| 2 | Cleaning of all filters and strainers |
| 3 | Pre commissioning pressurization fans |
| 4 | Water circulation in system and drain (dynamic) |
| 5 | Refilling of water and witness by Engineer |
| 6 | Prelim. Adjusting and balancing of valves |
| 7 | Test and commission chilled water pumps |
| 8 | Test and commission heat exchangers |
| 9 | Test and commission pressurization fans |
| 10 | Test and commission FAHU's |
| 11 | Test and commission floor FCU's |
| 12 | FAHU air balancing |
| 13 | Toilet exhaust / kitchen exhaust air balancing |
| 14 | Air balancing supply & return grilles |
| 15 | FCU water balancing |
| 16 | Balancing air from pressurization fans |
| 17 | Water balancing of FAHU's |
| 18 | Main System Valves adjustment |

# SECTION 2

## MEP ACTIVITY FLOW DIAGRAM

The main concept of develop the activity flow diagram to highlight the sequence of MEP activities installation in multistory buildings & towers. This flow diagram is very useful while preparing construction schedule.

The activity flow diagram deals only to get the idea of MEP activity overall system. For more detail i.e. MEP interface with civil activity and production rate refer the relevant section.

The flow diagram includes all drawing & material submittal and approval as well as material procurement and delivery sequence.

Third party involvement i.e. authority requirement, inspection and approval also include in the flow diagram.

List of Figures:-

**Material Submittal** — Material approved → **Material Procurement**

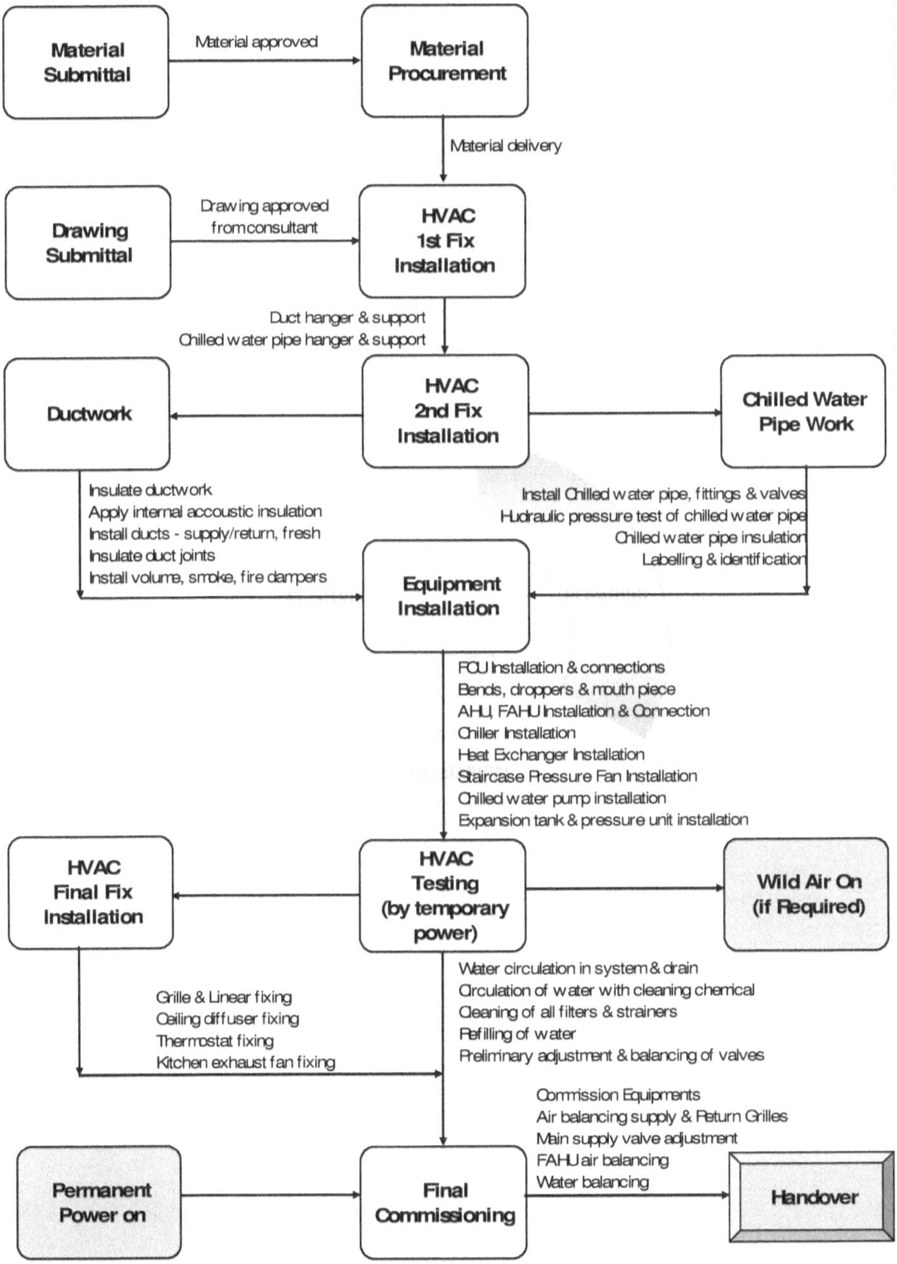

Material delivery

**Drawing Submittal** — Drawing approved from consultant → **HVAC 1st Fix Installation**

Duct hanger & support
Chilled water pipe hanger & support

**Ductwork** ← **HVAC 2nd Fix Installation** → **Chilled Water Pipe Work**

Insulate ductwork
Apply internal accoustic insulation
Install ducts - supply/return, fresh
Insulate duct joints
Install volume, smoke, fire dampers

Install Chilled water pipe, fittings & valves
Hudraulic pressure test of chilled water pipe
Chilled water pipe insulation
Labelling & identification

**Equipment Installation**

FCU Installation & connections
Bends, droppers & mouth piece
AHU, FAHU Installation & Connection
Chiller Installation
Heat Exchanger Installation
Staircase Pressure Fan Installation
Chilled water pump installation
Expansion tank & pressure unit installation

**HVAC Final Fix Installation** ← **HVAC Testing (by temporary power)** → **Wild Air On (if Required)**

Grille & Linear fixing
Ceiling diffuser fixing
Thermostat fixing
Kitchen exhaust fan fixing

Water circulation in system & drain
Circulation of water with cleaning chemical
Cleaning of all filters & strainers
Refilling of water
Preliminary adjustment & balancing of valves

Commission Equipments
Air balancing supply & Return Grilles
Main supply valve adjustment
FAHU air balancing
Water balancing

**Permanent Power on** → **Final Commissioning** → **Handover**

## *HVAC ACTIVITY FLOW DIAGRAM*

21

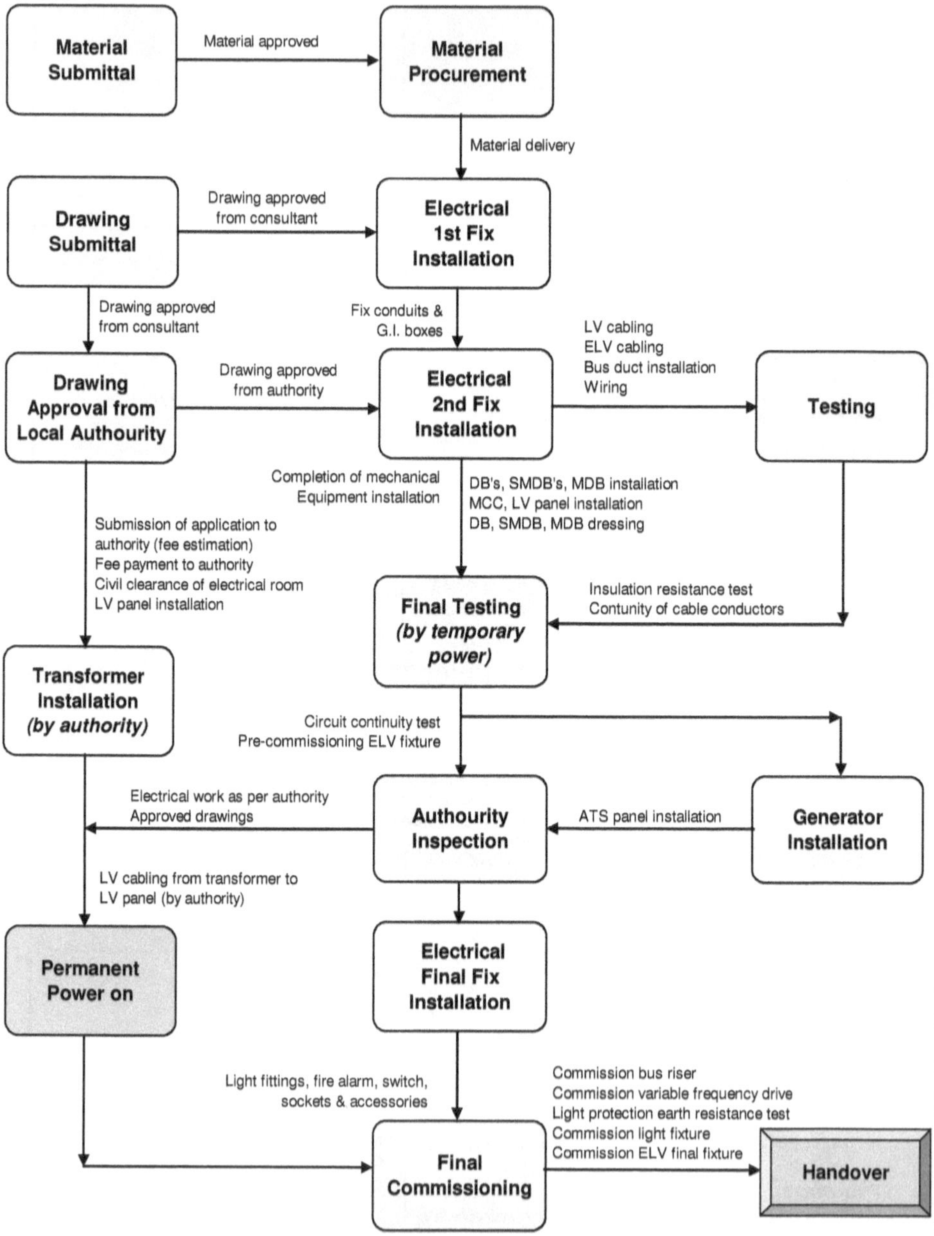

| | |
|---|---|
| PLANNING DEPARTMENT | |
| Rev.: 0 | **MEP PLANNING MANUAL** |
| Date of Issue: Feb 2007 | |
| Doc No.: MEP-01-07 | |

**Material Submittal** → *Material approved* → **Material Procurement**

*Material delivery*

**Drawing Submittal** → *Drawing approved from consultant* → **Electrical 1st Fix Installation**

*Drawing approved from consultant*

*Fix conduits & G.I. boxes*

*LV cabling*
*ELV cabling*
*Bus duct installation*
*Wiring*

**Drawing Approval from Local Authourity** → *Drawing approved from authority* → **Electrical 2nd Fix Installation** → **Testing**

*Completion of mechanical Equipment installation*

*DB's, SMDB's, MDB installation*
*MCC, LV panel installation*
*DB, SMDB, MDB dressing*

*Submission of application to authority (fee estimation)*
*Fee payment to authority*
*Civil clearance of electrical room*
*LV panel installation*

*Insulation resistance test*
*Contunity of cable conductors*

**Final Testing (by temporary power)**

**Transformer Installation (by authority)**

*Circuit continuity test*
*Pre-commissioning ELV fixture*

*Electrical work as per authority Approved drawings*

**Authourity Inspection** → *ATS panel installation* → **Generator Installation**

*LV cabling from transformer to LV panel (by authority)*

**Permanent Power on**

**Electrical Final Fix Installation**

*Light fittings, fire alarm, switch, sockets & accessories*

*Commission bus riser*
*Commission variable frequency drive*
*Light protection earth resistance test*
*Commission light fixture*
*Commission ELV final fixture*

**Final Commissioning** → **Handover**

## *ELECTRICAL ACTIVITY FLOW DIAGRAM*

```
┌──────────────┐  Material approved   ┌──────────────┐
│   Material   │ ───────────────────▶ │   Material   │
│   Submittal  │                      │  Procurement │
└──────────────┘                      └──────────────┘
                                              │
                                              │ Material delivery
                                              ▼
┌──────────────┐  Drawing approved    ┌──────────────┐
│   Drawing    │  from consultant     │   Plumbing   │
│   Submittal  │ ───────────────────▶ │   1st Fix    │
└──────────────┘                      │ Installation │
                                      └──────────────┘
```

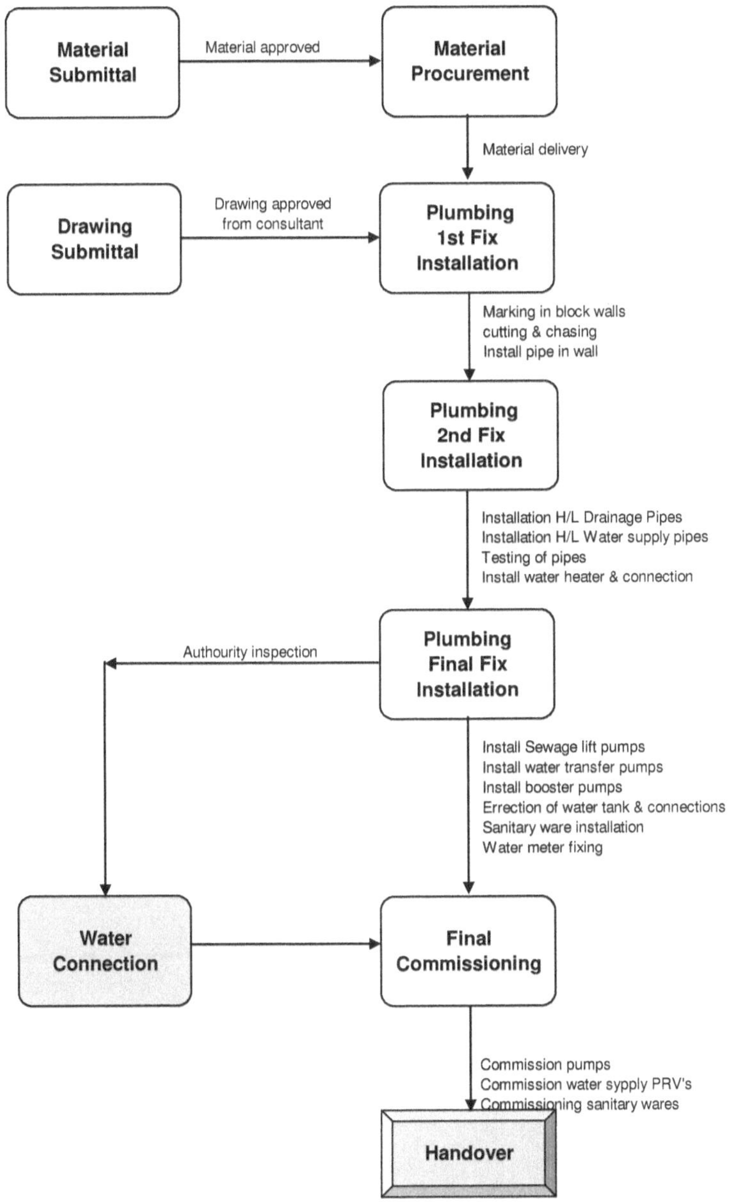

Marking in block walls
cutting & chasing
Install pipe in wall

**Plumbing
2nd Fix
Installation**

Installation H/L Drainage Pipes
Installation H/L Water supply pipes
Testing of pipes
Install water heater & connection

Authourity inspection ◀─── **Plumbing
Final Fix
Installation**

Install Sewage lift pumps
Install water transfer pumps
Install booster pumps
Errection of water tank & connections
Sanitary ware installation
Water meter fixing

**Water
Connection** ───▶ **Final
Commissioning**

Commission pumps
Commission water sypply PRV's
Commissioning sanitary wares

**Handover**

## *PLUMBING ACTIVITY FLOW DIAGRAM*

```
┌──────────────┐   Material approved   ┌──────────────┐
│   Material   │──────────────────────▶│   Material   │
│  Submittal   │                       │ Procurement  │
└──────────────┘                       └──────────────┘
                                              │
                                              │ Material delivery
                                              ▼
┌──────────────┐  Drawing approved     ┌──────────────┐
│   Drawing    │  from consultant      │ Fire Fighting│
│  Submittal   │──────────────────────▶│   1st Fix    │
└──────────────┘                       │ Installation │
        │                              └──────────────┘
        │ Drawing approved                    │
        │ from consultant       Marking, hangers & supports
        ▼                                     ▼
┌──────────────┐  Drawing approved     ┌──────────────┐  Install Fire Fighting Pipe Header
│   Drawing    │  from civil defence   │ Fire Fighting│  Install sprinklet pipelines & drops
│ Approval from│──────────────────────▶│   2nd Fix    │──────────────▶
│ Civil Defence│                       │ Installation │
└──────────────┘                       └──────────────┘
```

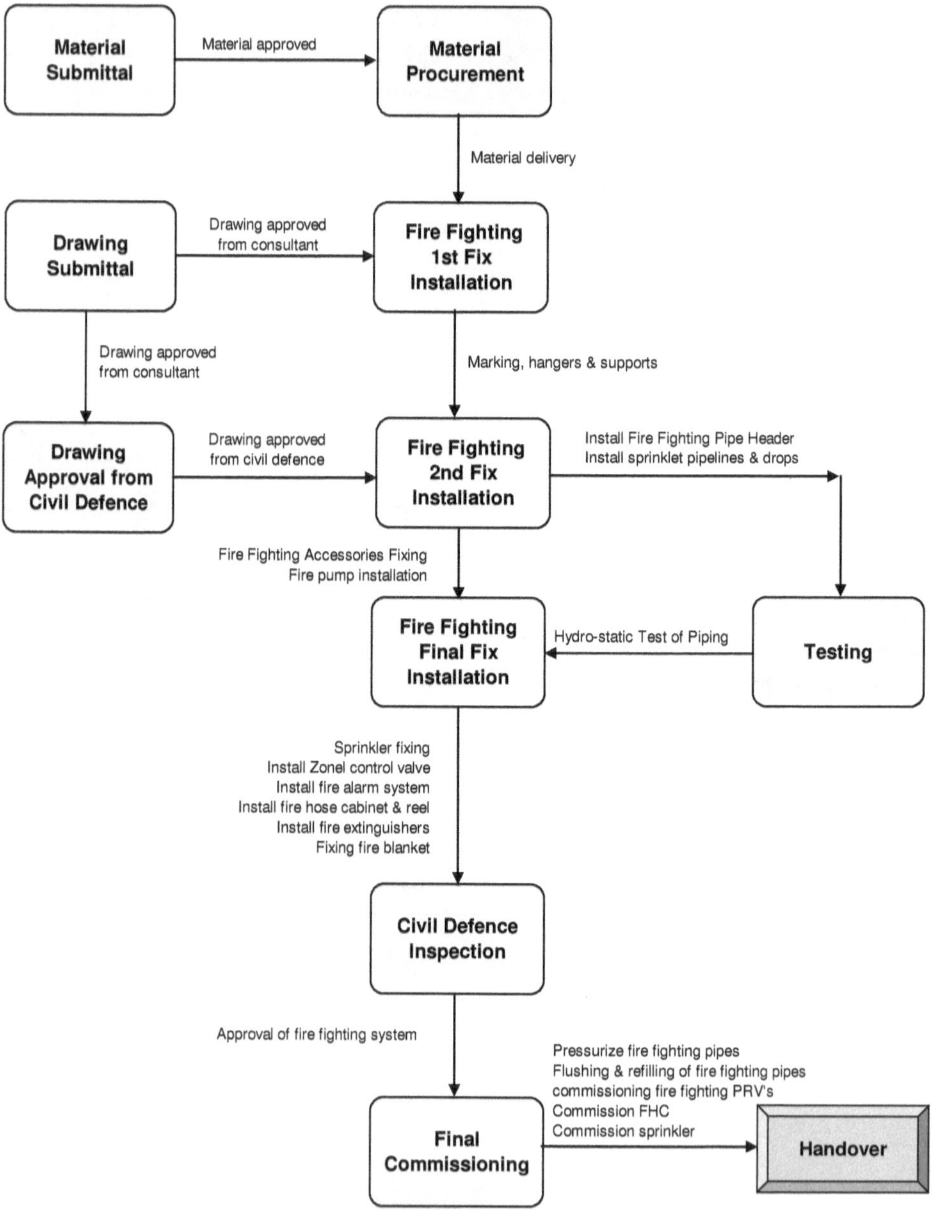

Fire Fighting Accessories Fixing
Fire pump installation

```
┌──────────────┐  Hydro-static Test of Piping   ┌──────────┐
│ Fire Fighting│◀───────────────────────────────│ Testing  │
│   Final Fix  │                                └──────────┘
│ Installation │
└──────────────┘
```

Sprinkler fixing
Install Zonel control valve
Install fire alarm system
Install fire hose cabinet & reel
Install fire extinguishers
Fixing fire blanket

```
┌──────────────┐
│ Civil Defence│
│  Inspection  │
└──────────────┘
```

Approval of fire fighting system

Pressurize fire fighting pipes
Flushing & refilling of fire fighting pipes
commissioning fire fighting PRV's
Commission FHC
Commission sprinkler

```
┌──────────────┐                  ┌──────────┐
│    Final     │─────────────────▶│ Handover │
│Commissioning │                  └──────────┘
└──────────────┘
```

## *FIRE FIGHTING ACTIVITY FLOW DIAGRAM*

# SECTION 3

# MEP AND CIVIL INTERFACE - GRAPHICAL

For better project monitoring and control it's very necessary to prepare the detailed program. To prepare the construction program without detailed MEP activity is not a good practice because lots of or we can say all civil finishing activities are directly related to MEP activities. For example without completing the high level piping & equipment installation civil contractor can't start their false ceiling work.

By the help of MEP and civil interface chart planning engineer can prepare the detailed integrated construction program in the initial stage for easy monitoring and controlling the project. Each service has separate chart and in the graphical mode for easy understanding.

This covers all MEP activity linked with the other civil activities, their relationship and their sequences in an engineering detailed manner. By the help of this activity relationship we can prepare more realistic construction program and track progress easily.

The MEP and civil interface chart has been made after several discussions with HEE Professional staff and the MEP sub-contractors.

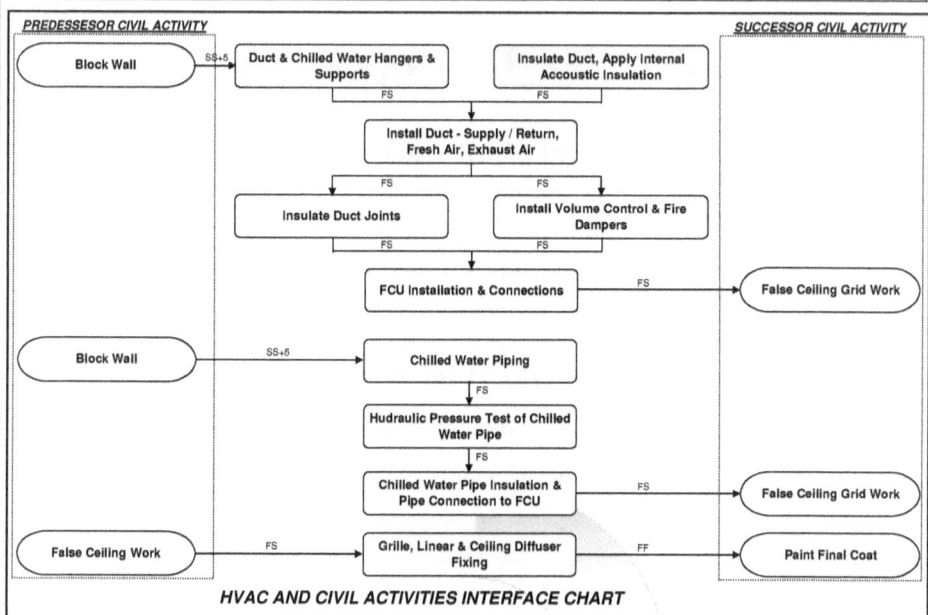

**HVAC AND CIVIL ACTIVITIES INTERFACE CHART**

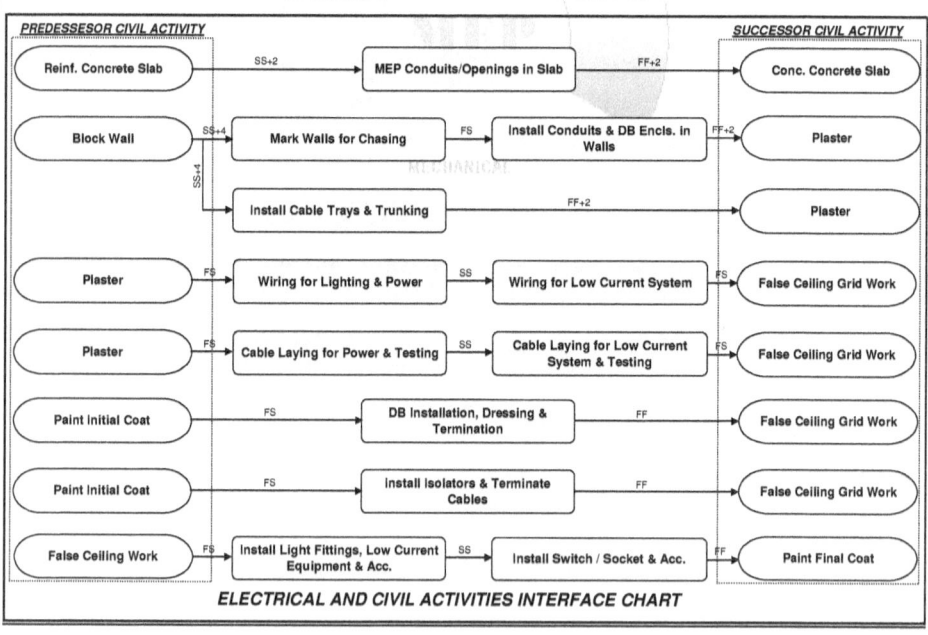

**ELECTRICAL AND CIVIL ACTIVITIES INTERFACE CHART**

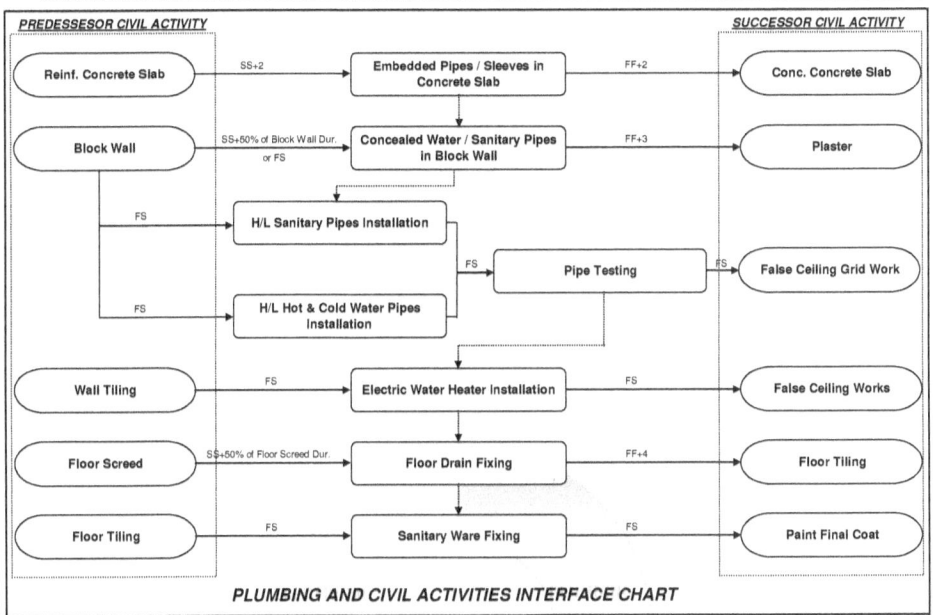

**PLUMBING AND CIVIL ACTIVITIES INTERFACE CHART**

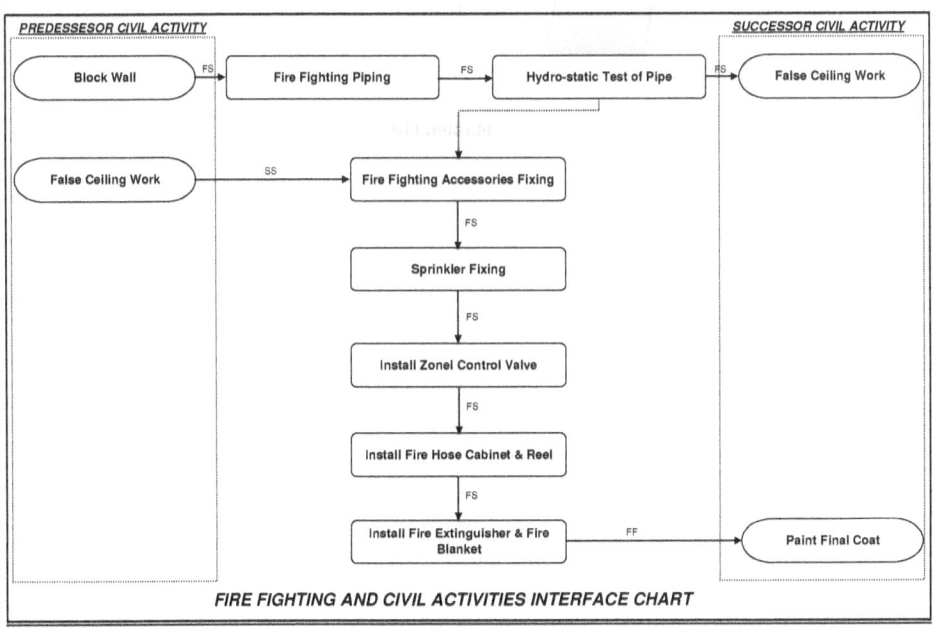

**FIRE FIGHTING AND CIVIL ACTIVITIES INTERFACE CHART**

# SECTION 4

## MEP AND CIVIL INTERFACE – TABULAR

For better project monitoring and control it's very necessary to prepare the detailed program. To prepare the construction program without detailed MEP activity is not a good practice because lots of or we can say all civil finishing activities are directly related to MEP activities. For example without completing the high level piping & equipment installation civil contractor can't start their false ceiling work.

By the help of MEP and civil interface chart planning engineer can prepare the detailed integrated construction program in the initial stage for easy monitoring and controlling the project. Each service has separate chart and in the graphical mode for easy understanding.

This covers all MEP activity linked with the other civil activities, their relationship and their sequences in an engineering detailed manner. By the help of this activity relationship we can prepare more realistic construction program and track progress easily.

The MEP and civil interface chart has been made after several discussions with HEE Professional staff and the MEP sub-contractors.

## Typical Multistoried Tower Programme Relationships - MEP WORK

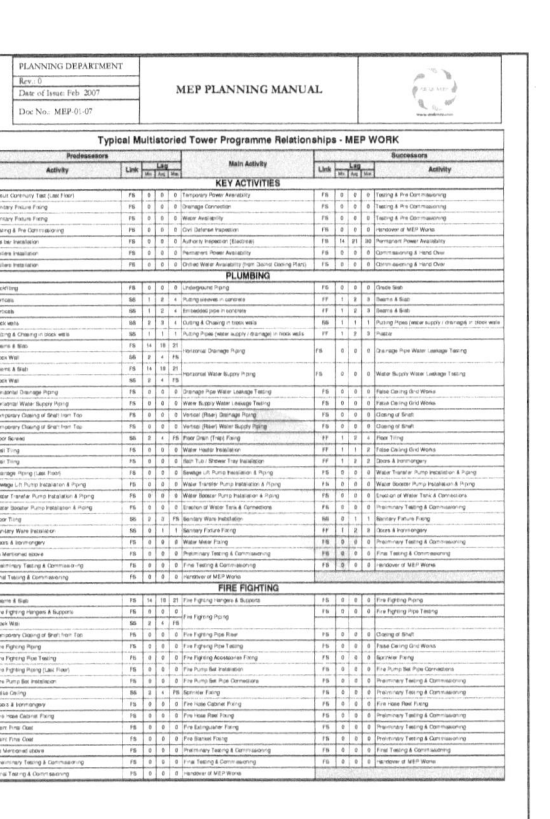

| Predecessors | | | | | Main Activity | Successors | | | | |
|---|---|---|---|---|---|---|---|---|---|---|
| Activity | Link | Min | Avg | Max | | Link | Min | Avg | Max | Activity |
| **KEY ACTIVITIES** | | | | | | | | | | |
| Circuit Continuity Test (Last Floor) | FS | 0 | 0 | 0 | Temporary Power Availability | FS | 0 | 0 | 0 | Testing & Pre Commissioning |
| Sanitary Fixture Fixing | FS | 0 | 0 | 0 | Drainage Connection | FS | 0 | 0 | 0 | Testing & Pre Commissioning |
| Sanitary Fixture Fixing | FS | 0 | 0 | 0 | Water Availability | FS | 0 | 0 | 0 | Testing & Pre Commissioning |
| Testing & Pre Commissioning | FS | 0 | 0 | 0 | Civil Defense Inspection | FS | 0 | 0 | 0 | Handover of MEP Works |
| Builder Installation | FS | 0 | 0 | 0 | Authority Inspection (Electrical) | FS | 14 | 21 | 30 | Permanent Power Availability |
| Chillers Installation | FS | 0 | 0 | 0 | Permanent Power Availability | FS | 0 | 0 | 0 | Commissioning & Hand Over |
| Chillers Installation | FS | 0 | 0 | 0 | Chilled Water Availability (from District Cooling Plant) | FS | 0 | 0 | 0 | Commissioning & Hand Over |
| **PLUMBING** | | | | | | | | | | |
| Backfilling | FS | 0 | 0 | 0 | Underground Piping | FS | 0 | 0 | 0 | Grade Slab |
| Verticals | SS | 1 | 3 | 4 | Puting sleeves in concrete | FF | 1 | 3 | 3 | Beams & Slab |
| Verticals | SS | 1 | 2 | 4 | Embedded pipe in concrete | FF | 1 | 2 | 3 | Beams & Slab |
| Block walls | SS | 0 | 3 | 4 | Cutting & Chasing in block walls | SS | 1 | 1 | 1 | Puting Pipes (water supply / drainage) in block walls |
| Cutting & Chasing in block walls | SS | 1 | 1 | 1 | Puting Pipes (water supply / drainage) in block walls | FF | 1 | 3 | 3 | Plaster |
| Beams & Slab | FS | 14 | 18 | 21 | Horizontal Drainage Piping | FS | 0 | 0 | 0 | Drainage Pipe Water Leakage Testing |
| Block Wall | SS | 4 | 4 | 15 | | | | | | |
| Beams & Slab | FS | 14 | 18 | 21 | Horizontal Water Supply Piping | FS | 0 | 0 | 0 | Water Supply Water Leakage Testing |
| Block Wall | SS | 2 | 4 | FS | | | | | | |
| Horizontal Drainage Piping | FS | 0 | 0 | 0 | Drainage Pipe Water Leakage Testing | FS | 0 | 0 | 0 | False Ceiling Grid Works |
| Horizontal Water Supply Piping | FS | 0 | 0 | 0 | Water Supply Water Leakage Testing | FS | 0 | 0 | 0 | False Ceiling Grid Works |
| Temporary Closing of Shaft from Top | FS | 0 | 0 | 0 | Vertical (Riser) Drainage Piping | FS | 0 | 0 | 0 | Closing of Shaft |
| Temporary Closing of Shaft from Top | FS | 0 | 0 | 0 | Vertical (Riser) Water Supply Piping | FS | 0 | 0 | 0 | Closing of Shaft |
| Floor Screed | SS | 3 | 4 | FS | Floor Drain (Trap) Fixing | FF | 1 | 0 | 4 | Floor Tiling |
| Wall Tiling | FS | 0 | 0 | 0 | Water Heater Installation | FF | 1 | 1 | 2 | False Ceiling Grid Works |
| Wall Tiling | FS | 0 | 0 | 0 | Bath Tub / Shower Tray Installation | FF | 1 | 2 | 2 | Doors & Ironmongery |
| Drainage Piping (Last Floor) | FS | 0 | 0 | 0 | Sewage Lift Pump Installation & Piping | FS | 0 | 0 | 0 | Water Transfer Pump Installation & Piping |
| Sewage Lift Pump Installation & Piping | FS | 0 | 0 | 0 | Water Transfer Pump Installation & Piping | FS | 0 | 0 | 0 | Water Booster Pump Installation & Piping |
| Water Transfer Pump Installation & Piping | FS | 0 | 0 | 0 | Water Booster Pump Installation & Piping | FS | 0 | 0 | 0 | Erection of Water Tank & Connections |
| Water Booster Pump Installation & Piping | FS | 0 | 0 | 0 | Erection of Water Tank & Connections | FS | 0 | 0 | 0 | Preliminary Testing & Commissioning |
| Floor Tiling | SS | 2 | 0 | FS | Sanitary Ware Installation | SS | 0 | 1 | 1 | Sanitary Fixture Fixing |
| Sanitary Ware Installation | SS | 0 | 1 | 1 | Sanitary Fixture Fixing | FF | 1 | 2 | 3 | Doors & Ironmongery |
| Doors & Ironmongery | FS | 0 | 0 | 0 | Water Meter Fixing | FS | 0 | 0 | 0 | Preliminary Testing & Commissioning |
| As Mentioned above | FS | 0 | 0 | 0 | Preliminary Testing & Commissioning | FS | 0 | 0 | 0 | Final Testing & Commissioning |
| Preliminary Testing & Commissioning | FS | 0 | 0 | 0 | Final Testing & Commissioning | FS | 0 | 0 | 0 | Handover of MEP Works |
| Final Testing & Commissioning | FS | 0 | 0 | 0 | Handover of MEP Works | | | | | |
| **FIRE FIGHTING** | | | | | | | | | | |
| Beams & Slab | FS | 14 | 18 | 21 | Fire Fighting Hangers & Supports | FS | 0 | 0 | 0 | Fire Fighting Piping |
| Fire Fighting Hangers & Supports | FS | 0 | 0 | 0 | Fire Fighting Piping | FS | 0 | 0 | 0 | Fire Fighting Pipe Testing |
| Block Wall | SS | 3 | 4 | FS | | | | | | |
| Temporary Closing of Shaft from Top | FS | 0 | 0 | 0 | Fire Fighting Pipe Riser | FS | 0 | 0 | 0 | Closing of Shaft |
| Fire Fighting Piping | FS | 0 | 0 | 0 | Fire Fighting Pipe Testing | FS | 0 | 0 | 0 | False Ceiling Grid Works |
| Fire Fighting Pipe Testing | FS | 0 | 0 | 0 | Fire Fighting Accessories Fixing | FS | 0 | 0 | 0 | Sprinkler Fixing |
| Fire Fighting Piping (Last Floor) | FS | 0 | 0 | 0 | Fire Pump Set Installation | FS | 0 | 0 | 0 | Fire Pump Set Pipe Connections |
| Fire Pump Set Installation | FS | 0 | 0 | 0 | Fire Pump Set Pipe Connections | FS | 0 | 0 | 0 | Preliminary Testing & Commissioning |
| False Ceiling | SS | 0 | 4 | FS | Sprinkler Fixing | FS | 0 | 0 | 0 | Preliminary Testing & Commissioning |
| Doors & Ironmongery | FS | 0 | 0 | 0 | Fire Hose Cabinet Fixing | FS | 0 | 0 | 0 | Fire Hose Reel Fixing |
| Fire Hose Cabinet Fixing | FS | 0 | 0 | 0 | Fire Hose Reel Fixing | FS | 0 | 0 | 0 | Preliminary Testing & Commissioning |
| Paint Final Coat | FS | 0 | 0 | 0 | Fire Extinguisher Fixing | FS | 0 | 0 | 0 | Preliminary Testing & Commissioning |
| Paint Final Coat | FS | 0 | 0 | 0 | Fire Blanket Fixing | FS | 0 | 0 | 0 | Preliminary Testing & Commissioning |
| As Mentioned above | FS | 0 | 0 | 0 | Preliminary Testing & Commissioning | FS | 0 | 0 | 0 | Final Testing & Commissioning |
| Preliminary Testing & Commissioning | FS | 0 | 0 | 0 | Final Testing & Commissioning | FS | 0 | 0 | 0 | Handover of MEP Works |
| Final Testing & Commissioning | FS | 0 | 0 | 0 | Handover of MEP Works | | | | | |

### Typical Multistoried Tower Programme Relationships - MEP WORK

| Predecessors | | | | | Main Activity | Successors | | | | |
|---|---|---|---|---|---|---|---|---|---|---|
| Activity | Link | Lag | | | | Link | Lag | | | Activity |
| | | Mn | Avg | Mrx | | | Mn | Avg | Mrx | |
| | | | | | **HVAC** | | | | | |
| Beams & floor | FS | 14 | 18 | 21 | Duct Hangers & Supports | FS | 0 | 0 | 0 | Ducting Supply, return, Fresh & Exhaust |
| Beams & floor | FS | 14 | 18 | 21 | Chilled Water Pipe Hangers & Supports | FS | 0 | 0 | 0 | Chilled Water Piping |
| Duct hangers & Supports | FS | 0 | 0 | 0 | Ducting Supply, return, Fresh & Exhaust | FF | 1 | 2 | 3 | Volume, Fire & Smoke Damper Fixing |
| Ducting Supply, return, Fresh & Exhaust | FF | 1 | 2 | 3 | Volume, Fire & Smoke Damper Fixing | FF | 1 | 2 | 3 | Duct Insulation |
| Temporary Closing of Shaft from Top | FS | 0 | 0 | 0 | Duct Risers | FS | 0 | 0 | 0 | Closing of Shaft |
| Volume, Fire & Smoke Damper Fixing | FF | 0 | 0 | 0 | Duct Insulation | FS | 0 | 0 | 0 | FCU Installation |
| Chilled Water Pipe Hangers & Supports | FS | 0 | 0 | 0 | Chilled Water Piping | FS | 0 | 0 | 0 | Chilled Water Pipe Test |
| Temporary Closing of Shaft from Top | FS | 0 | 0 | 0 | Chilled Water Pipe Riser | FS | 0 | 0 | 0 | Closing of Shaft |
| Chilled Water Piping | FS | 0 | 0 | 0 | Chilled Water Pipe Test | FS | 0 | 0 | 0 | Chilled Water Pipe Insulation |
| Chilled Water Pipe Test | FS | 0 | 0 | 0 | Chilled Water Pipe Insulation | FS | 0 | 0 | 0 | FCU Installation |
| Chilled Water Pipe Insulation | FS | 0 | 0 | 0 | FCU Installation | FF | 1 | 1 | 2 | False Ceiling Grid Works |
| False Ceiling Grid Works | FS | 0 | 0 | 0 | Bends, Droppers & Mouth Piece | FS | 0 | 0 | 0 | MEP Clearance for False Ceiling Closure |
| Bends, Droppers & Mouth Piece | FS | 0 | 0 | 0 | MEP Clearance for False Ceiling Closure | FS | 0 | 0 | 0 | False Ceiling Work |
| Roof Waterproofing | FS | 0 | 0 | 0 | Chillers Installation | FS | 0 | 0 | 0 | Staircase Pressure Fan Installation |
| Doors & Ironmongery | FS | 0 | 0 | 0 | Heat Exchanger Installation | FS | 0 | 0 | 0 | Preliminary Testing & Commissioning |
| Chillers Installation | FS | 0 | 0 | 0 | Staircase Pressure Fan Installation | FS | 0 | 0 | 0 | Garbage Exhaust Fan Installation |
| Staircase Pressure Fan Installation | FS | 0 | 0 | 0 | Garbage Exhaust Fan Installation | FS | 0 | 0 | 0 | Preliminary Testing & Commissioning |
| Doors & Ironmongery | FS | 0 | 0 | 0 | Chilled Water Pump Installation | FS | 0 | 0 | 0 | Expansion Tank & Pressure Unit Installation |
| Chilled Water Pump Installation | FS | 0 | 0 | 0 | Expansion Tank & Pressure Unit Installation | FS | 0 | 0 | 0 | Air Scrubbers Installation |
| Expansion Tank & Pressure Unit Installation | FS | 0 | 0 | 0 | Air Scrubbers Installation | FS | 0 | 0 | 0 | Chemical Dosing Plant Installation |
| Air Scrubbers Installation | FS | 0 | 0 | 0 | Chemical Dosing Plant Installation | FS | 0 | 0 | 0 | AHU Installation |
| Chemical Dosing Plant Installation | FS | 0 | 0 | 0 | AHU Installation | FS | 0 | 0 | 0 | Preliminary Testing & Commissioning |
| Doors & Ironmongery | FS | 0 | 0 | 0 | Kitchen Exhaust Fan Fixing | FS | 0 | 0 | 0 | Preliminary Testing & Commissioning |
| Bends, Droppers & Mouth Piece | FS | 0 | 0 | 0 | Ceiling Diffuser Fixing | FS | 0 | 0 | 0 | Grille & Linear Fixing |
| Ceiling Diffuser Fixing | FS | 0 | 0 | 0 | Grille & Linear Fixing | FS | 0 | 0 | 0 | Thermostat Fixing |
| Grille & Linear Fixing | FS | 0 | 0 | 0 | Thermostat Fixing | FS | 0 | 0 | 0 | Preliminary Testing & Commissioning |
| As Mentioned above | FS | 0 | 0 | 0 | Preliminary Testing & Commissioning | FS | 0 | 0 | 0 | Final Testing & Commissioning |
| Preliminary Testing & Commissioning | FS | 0 | 0 | 0 | Final Testing & Commissioning | FS | 0 | 0 | 0 | Handover of MEP Works |
| Final Testing & Commissioning | FS | 0 | 0 | 0 | Handover of MEP Works | | | | | |

| PLANNING DEPARTMENT | | |
|---|---|---|
| Rev. 0 | **MEP PLANNING MANUAL** |  |
| Date of Issue: Feb. 2007 | | www.mishraje.com |
| Doc No.: MEP-01-07 | | |

## Typical Multistoried Tower Programme Relationships - MEP WORK

| Predecessors | | | | | Main Activity | Successors | | | | |
|---|---|---|---|---|---|---|---|---|---|---|
| Activity | Link | Lag | | | | Link | Lag | | | Activity |
| | | Min | Avg | Max | | | Min | Avg | Max | |
| | | | | | **ELECTRICAL** | | | | | |
| Verticals | SS | 1 | 2 | 4 | Embedded Conduits in Concrete | FS | 0 | 0 | 0 | Grade Slab |
| Block Wall | SS | 2 | 3 | 4 | Marking on Block walls | SS | 1 | 2 | 0 | Cutting & Chasing in block walls |
| Marking on Block walls | SS | 1 | 2 | 0 | Cutting & Chasing in block walls | SS | 1 | 2 | 3 | Fix Conduits, GI Boxes in Block Walls |
| Cutting & Chasing in block walls | SS | 1 | 2 | 2 | Fix Conduits, GI Boxes in Block Walls | FF | 1 | 3 | 3 | Plaster |
| Plaster | FS | 0 | 0 | 0 | GI & PVC Surface Conducting | FF | 2 | 4 | 4 | False Ceiling Work |
| Cutting & Chasing in block walls | SS | 1 | 2 | 3 | Installation of Distribution Boards | SS | 1 | 2 | 0 | Installation of Sub-Main Boards |
| Installation of Distribution Boards | SS | 1 | 3 | 3 | Installation of Sub-Main Boards | FF | 1 | 2 | 3 | Plaster |
| Paint Initial Coat | FS | 0 | 0 | 0 | Installation of Main LV Panel | FF | 0 | 0 | 0 | Doors & Ironmongery |
| Block Wall | SS | 4 | 6 | FS | Erect Cable Tray | FF | 0 | 1 | 1 | Plaster |
| Plaster | FS | 0 | 0 | 0 | Cable Trunking | FS | 0 | 0 | 0 | Earthing System |
| Cable Trunking | FS | 0 | 0 | 0 | Earthing System | FS | 0 | 0 | 0 | LV Cabling |
| Earthing System | FS | 0 | 0 | 0 | LV Cabling | FS | 0 | 0 | 0 | ELV Cabling |
| LV Cabling | FS | 0 | 0 | 0 | ELV Cabling | FS | 0 | 0 | 0 | Main LV Cable Termination |
| ELV Cabling | FS | 0 | 0 | 0 | Main LV Cable Termination | FS | 0 | 0 | 0 | Sub-main Cable Termination |
| Main LV Cable Termination | FS | 0 | 0 | 0 | Sub-main Cable Termination | FS | 0 | 0 | 0 | Wiring for Lighting & Small Power |
| Sub-main Cable Termination | FS | 0 | 0 | 0 | Wiring for Lighting & Small Power | FS | 0 | 0 | 0 | Wiring for ELV System |
| Wiring for Lighting & Small Power | FS | 0 | 0 | 0 | Wiring for ELV System | FS | 0 | 0 | 0 | Wiring for Fire Fighting System |
| Wiring for ELV System | FS | 0 | 0 | 0 | Wiring for Fire Fighting System | FS | 0 | 0 | 0 | Circuit Continuity Test |
| Wiring for Fire Fighting System | FS | 0 | 0 | 0 | Circuit Continuity Test | FS | 0 | 0 | 0 | Ceiling Rose & Flexible Drops |
| Circuit Continuity Test | FS | 0 | 0 | 0 | Ceiling Rose & Flexible Drops | FF | 3 | 4 | 4 | False Ceiling Work |
| Ceiling Rose & Flexible Drops | FS | 0 | 0 | 0 | Distribution Boards Dressing | FF | 0 | 1 | 1 | Paint Final Coat |
| LV Cabling (Last Floor) | FS | 0 | 0 | 0 | Bus bar Installation | SS | 0 | 0 | 0 | Lightning Protection System |
| Bus bar Installation | SS | 0 | 0 | 0 | Lightning Protection System | FS | 0 | 0 | 0 | Transformer Installation |
| Lightning Protection System | FS | 0 | 0 | 0 | Transformer Installation | FS | 0 | 0 | 0 | Generator Installation |
| Transformer Installation | FS | 0 | 0 | 0 | Generator Installation | FS | 0 | 0 | 0 | ATS Panel Installation |
| Generator Installation | FS | 0 | 0 | 0 | ATS Panel Installation | FS | 0 | 0 | 0 | LV Cabling from Transformer to LV Panel |
| ATS Panel Installation | FS | 0 | 0 | 0 | LV Cabling from Transformer to LV Panel | FF | 1 | 2 | 2 | Paint Final Coat |
| Doors & Ironmongery | FS | 0 | 0 | 0 | Fixing Fire Alarm & Light Fittings | FS | 0 | 0 | 0 | Fixing Switches, Sockets & Accessories |
| Fixing Fire Alarm & Light Fittings | FS | 0 | 0 | 0 | Fixing Switches, Sockets & Accessories | FF | 1 | 2 | 2 | Paint Final Coat |
| As Mentioned above | FS | 0 | 0 | 0 | Preliminary Testing & Commissioning | FS | 0 | 0 | 0 | Final Testing & Commissioning |
| Preliminary Testing & Commissioning | FS | 0 | 0 | 0 | Final Testing & Commissioning | FS | 0 | 0 | 0 | Handover of MEP Works |
| Final Testing & Commissioning | FS | 0 | 0 | 0 | Handover of MEP Works | | | | | |

| PLANNING DEPARTMENT | |
| --- | --- |
| Rev.: 0 | |
| Date of Issue: Feb 2007 | **MEP PLANNING MANUAL** |
| Doc No.: MEP-01-07 | |

# SECTION 5

# MEP ACTIVITY PRODUCTION RATE

The production rate should be the planned average daily production rate based on available resource and the environment within which the work will be performed. The production rate include not only the productive time associated with a particular activity but also idle and non-productive time associated with the movement and setup of equipment, break times, expected equipment downtime for maintenance and repair, and other expected downtime. When estimating a production rate, resource availability and other factors affecting productive need to be considered.

This section covers the manpower requirement to complete each activity and the time required to complete. This production rate has been prepared by MEP Planning Engineer after detailed discussion with professional MEP contractors. This will help Planning Engineer to calculate the activity durations in a professional manner.

The production rate given in this section, on the other hand, is not a constant and can be affected by a number of factors both within and beyond the contractor's control.

PLANNING DEPARTMENT
Rev.: 0
Date of Issue: Feb 2007
Doc No.: MEP-01-07

**MEP PLANNING MANUAL**

www.arabmep.com

# PRODUCTION RATE FOR MEP ACTIVITY — **MECHANICAL**

| Description | Unit | Qty./day (Average) | Qty./hr (Average) | PL | PF | PW | DM | IN | EL | HL | Total Manpower Per Group |
|---|---|---|---|---|---|---|---|---|---|---|---|
| **FIRE FIGHTING PIPES** | | | | | | | | | | | |
| 25mm diameter black seamless steel pipe with threaded fittings | m | 30 | 3.75 | | 1 | | | | | 1 | 2 |
| 32mm diameter black seamless steel pipe with threaded fittings | m | 30 | 3.75 | | 1 | | | | | 1 | 2 |
| 40mm diameter black seamless steel pipe with threaded fittings | m | 30 | 3.75 | | 1 | | | | | 1 | 2 |
| 50mm diameter black seamless steel pipe with threaded fittings | m | 30 | 3.75 | | 1 | | | | | 1 | 2 |
| 65mm diameter black seamless steel pipe with grooved-end | m | 20 | 2.5 | | 1 | | | | | 1 | 2 |
| 80mm diameter black seamless steel pipe with grooved-end | m | 20 | 2.5 | | 1 | | | | | 2 | 3 |
| 100mm diameter black seamless steel pipe with grooved-end | m | 20 | 2.5 | | 1 | | | | | 2 | 3 |
| 125mm diameter black seamless steel pipe with grooved-end | m | 20 | 2.5 | | 1 | | | | | 2 | 3 |
| 150mm diameter black seamless steel pipe with grooved-end | m | 18 | 2.25 | | 2 | | | | | 4 | 6 |
| 200mm diameter black seamless steel pipe with grooved-end | m | 18 | 2.25 | | 2 | | | | | 4 | 6 |
| 250mm diameter black seamless steel pipe with grooved-end | m | 18 | 2.25 | | 2 | | | | | 4 | 6 |
| 300mm diameter black seamless steel pipe with grooved-end | m | 18 | 2.25 | | 2 | | | | | 4 | 6 |
| **HOT AND COLD WATER DISTRIBUTION PIPES** | | | | | | | | | | | |
| 20mm diameter C.P.V.C. pipes include fittings | m | 36 | 4.5 | 1 | | | | | | 1 | 2 |
| 25mm diameter C.P.V.C. pipes include fittings | m | 36 | 4.5 | 1 | | | | | | 1 | 2 |
| 32mm diameter C.P.V.C. pipes include fittings | m | 36 | 4.5 | 1 | | | | | | 1 | 2 |
| 40mm diameter C.P.V.C. pipes include fittings | m | 36 | 4.5 | 1 | | | | | | 1 | 2 |
| 50mm diameter C.P.V.C. pipes include fittings | m | 36 | 4.5 | 1 | | | | | | 1 | 2 |
| 63mm diameter C.P.V.C. pipes include fittings | m | 24 | 3 | 1 | | | | | | 2 | 3 |
| 75mm diameter C.P.V.C. pipes include fittings | m | 24 | 3 | 1 | | | | | | 2 | 3 |
| 90mm diameter C.P.V.C. pipes include fittings | m | 24 | 3 | 1 | | | | | | 2 | 3 |
| 110mm diameter C.P.V.C. pipes include fittings | m | 24 | 3 | 1 | | | | | | 2 | 3 |
| 125mm diameter C.P.V.C. pipes include fittings | m | 22 | 2.75 | 1 | | | | | | 2 | 3 |
| 160mm diameter C.P.V.C. pipes include fittings | m | 22 | 2.75 | 1 | | | | | | 2 | 3 |
| **SANITARY AND RAINWATER PIPES** | | | | | | | | | | | |
| 40mm diameter UPVC pipes with elastomere joint socket include fittings | m | 36 | 4.5 | 1 | | | | | | 1 | 2 |
| 50mm diameter UPVC pipes with elastomere joint socket include fittings | m | 36 | 4.5 | 1 | | | | | | 1 | 2 |
| 63mm diameter UPVC pipes with elastomere joint socket include fittings | m | 30 | 3.75 | 1 | | | | | | 1 | 2 |
| 75mm diameter UPVC pipes with elastomere joint socket include fittings | m | 30 | 3.75 | 1 | | | | | | 1 | 2 |
| 110mm diameter UPVC pipes with elastomere joint socket include fittings | m | 24 | 3 | 1 | | | | | | 2 | 3 |
| 125mm diameter UPVC pipes with elastomere joint socket include fittings | m | 24 | 3 | 1 | | | | | | 2 | 3 |
| 160mm diameter UPVC pipes with elastomere joint socket include fittings | m | 18 | 2.25 | 1 | | | | | | 2 | 3 |
| 200mm diameter UPVC pipes with elastomere joint socket include fittings | m | 18 | 2.25 | 1 | | | | | | 2 | 3 |
| 25mm diameter PVC condensate pipes with insulation | m | 30 | 3.75 | 1 | | | | 1 | | 1 | 3 |
| 32mm diameter PVC condensate pipes with insulation | m | 30 | 3.75 | 1 | | | | 1 | | 1 | 3 |
| 40mm diameter PVC condensate pipes with insulation | m | 30 | 3.75 | 1 | | | | 1 | | 1 | 3 |
| 50mm diameter PVC condensate pipes with insulation | m | 30 | 3.75 | 1 | | | | 1 | | 1 | 3 |
| 60mm diameter PVC condensate pipes with insulation | m | 20 | 2.5 | 1 | | | | 1 | | 1 | 3 |
| 110mm diameter PVC condensate pipes with insulation | m | 20 | 2.5 | 1 | | | | 1 | | 2 | 4 |

# PRODUCTION RATE FOR MEP ACTIVITY

## MECHANICAL

| Description | Unit | Qty./day (Average) | Qty./hr (Average) | Manpower Requirement Per Group | | | | | | | Total Manpower Per Group |
|---|---|---|---|---|---|---|---|---|---|---|---|
| | | | | PL | PF | PW | DM | IN | EL | HL | |
| **HYDRONIC PIPES (CHILLED WATER PIPES)** | | | | | | | | | | | |
| 15mm diameter seamless black steel pipe include fittings | m | 30 | 3.75 | | 1 | | | | | 1 | 2 |
| 20mm diameter seamless black steel pipe include fittings | m | 30 | 3.75 | | 1 | | | | | 1 | 2 |
| 25mm diameter seamless black steel pipe include fittings | m | 30 | 3.75 | | 1 | | | | | 1 | 2 |
| 32mm diameter seamless black steel pipe include fittings | m | 30 | 3.75 | | 1 | | | | | 1 | 2 |
| 40mm diameter seamless black steel pipe include fittings | m | 30 | 3.75 | | 1 | | | | | 1 | 2 |
| 50mm diameter seamless black steel pipe include fittings | m | 30 | 3.75 | | 1 | | | | | 1 | 2 |
| 65mm diameter seamless black steel pipe include fittings | m | 20 | 2.5 | | 1 | | | | | 1 | 2 |
| 80mm diameter seamless black steel pipe include fittings | m | 20 | 2.5 | | 1 | | | | | 1 | 2 |
| 100mm diameter seamless black steel pipe include fittings | m | 20 | 2.5 | | 1 | | | | | 2 | 3 |
| 125mm diameter seamless black steel pipe include fittings | m | 20 | 2.5 | | 1 | | | | | 2 | 3 |
| 150mm diameter seamless black steel pipe include fittings | m | 20 | 2.5 | | 1 | | | | | 2 | 3 |
| 200mm diameter seamless black steel pipe include fittings | m | 18 | 2.25 | | 2 | | | | | 4 | 6 |
| 250mm diameter seamless black steel pipe include fittings | m | 18 | 2.25 | | 2 | | | | | 4 | 6 |
| 300mm diameter seamless black steel pipe include fittings | m | 18 | 2.25 | | 2 | | | | | 4 | 6 |
| **HYDRONIC PIPES INSULATION (CHILLED WATER PIPES)** | | | | | | | | | | | |
| 15mm diameter pipe insulation | m | 50 | 6.25 | | | | | 1 | | 1 | 2 |
| 20mm diameter pipe insulation | m | 50 | 6.25 | | | | | 1 | | 1 | 2 |
| 25mm diameter pipe insulation | m | 50 | 6.25 | | | | | 1 | | 1 | 2 |
| 32mm diameter pipe insulation | m | 50 | 6.25 | | | | | 1 | | 1 | 2 |
| 40mm diameter pipe insulation | m | 50 | 6.25 | | | | | 1 | | 1 | 2 |
| 50mm diameter pipe insulation | m | 50 | 6.25 | | | | | 1 | | 1 | 2 |
| 65mm diameter pipe insulation | m | 30 | 3.75 | | | | | 1 | | 1 | 2 |
| 80mm diameter pipe insulation | m | 30 | 3.75 | | | | | 1 | | 1 | 2 |
| 100mm diameter pipe insulation | m | 30 | 3.75 | | | | | 1 | | 1 | 2 |
| 125mm diameter pipe insulation | m | 20 | 2.5 | | | | | 1 | | 1 | 2 |
| 150mm diameter pipe insulation | m | 20 | 2.5 | | | | | 1 | | 1 | 2 |
| 200mm diameter pipe insulation | m | 20 | 2.5 | | | | | 1 | | 1 | 2 |
| 250mm diameter pipe insulation | m | 16 | 2 | | | | | 1 | | 1 | 2 |
| 300mm diameter pipe insulation | m | 16 | 2 | | | | | 1 | | 1 | 2 |

ARAB MEP

www.arabmep.com

# PRODUCTION RATE FOR MEP ACTIVITY | MECHANICAL

| Description | Unit | Qty./day (Average) | Qty./hr (Average) | PL | PF | PW | DM | IN | EL | HL | Total Manpower Per Group |
|---|---|---|---|---|---|---|---|---|---|---|---|
| **PLUMBING EQUIPMENT** | | | | | | | | | | | |
| Water distribution pumps installation (one set) | nos | 1 | 0.125 | | 2 | | | | | 2 | 4 |
| Booster pumps installation | nos | 1 | 0.125 | | 2 | | | | | 2 | 4 |
| Sewage lift pumps installaation | nos | 1 | 0.125 | | 2 | | | | | 2 | 4 |
| Electric water heater installation | nos | 2 | 0.25 | 1 | | | | | | 2 | 3 |
| Bath tub installation | nos | 1 | 0.125 | 1 | | | | | | 2 | 3 |
| Shower tray installation | nos | 1 | 0.125 | 1 | | | | | | 2 | 3 |
| Water closet installation | nos | 1 | 0.125 | 1 | | | | | | 1 | 2 |
| Bidet installation | nos | 1 | 0.13 | 1 | | | | | | 1 | 2 |
| Wash basin installation | nos | 1 | 0.13 | 1 | | | | | | 1 | 2 |
| Kitchen sink installation | nos | 1 | 0.125 | 1 | | | | | | 1 | 2 |
| **FIRE FIGHTING EQUIPMENT** | | | | | | | | | | | |
| Fire water pumps installation | nos | 1 | 0.13 | | 2 | | | | | 2 | 4 |
| Fire hose cabinet installaation | nos | 1 | 0.13 | | 2 | | | | | 2 | 4 |
| Sprinkler fixing | nos | 30 | 3.75 | | 1 | | | | | 1 | 2 |
| **HVAC EQUIPMENT** | | | | | | | | | | | |
| Fan coil unit (FCU) installation | nos | 4 | 0.50 | | 2 | | | | | 2 | 4 |
| Air handling unit (AHU) installation | nos | 1 | 0.13 | | 4 | | | | | 4 | 8 |
| Fresh air handling unit (FAHU) installation | nos | 1 | 0.13 | | 4 | | | | | 4 | 8 |
| Chilled water pump installation | nos | 1 | 0.13 | | 2 | | | | | 2 | 4 |
| **DUCTING** | | | | | | | | | | | |
| Duct installation | sq.ft | 60 | 7.50 | | | | 1 | | | 1 | 2 |
| Duct insulation | sq.ft | 60 | 7.50 | | | | | 1 | | 1 | 2 |
| Duct connection to FCU | nos | 4 | 0.50 | | 1 | | | | | 1 | 2 |
| Duct connection to AHU | nos | 1 | 0.13 | | 4 | | | | | 4 | 8 |
| Duct connection to FAHU | nos | 1 | 0.13 | | 4 | | | | | 4 | 8 |
| Chilled water pipe connection to FCU / AHU / FAHU | nos | 1 | 0.13 | | 4 | | | | | 4 | 8 |

| | |
|---|---|
| PLANNING DEPARTMENT | |
| Rev.: 0 | **MEP PLANNING MANUAL** |
| Date of Issue: Feb 2007 | |
| Doc No.: MEP-01-07 | |

# SECTION 6

## MATERIAL PROCUREMENT

Procurement activities are those activities that are necessary to ensure that needed labor, material, and equipment are available at the project site when needed. Procurement activities are a part of the construction process and must be scheduled as such. An easy and effective means of arranging procurement activities in a construction schedule is the use of procurement ladder mentioned below.

➢ Material Submit  for Review

➢ Engineer Approval of Material

➢ Procurement (Order Material)

➢ Fabricate & Ship Material

➢ Receive Material

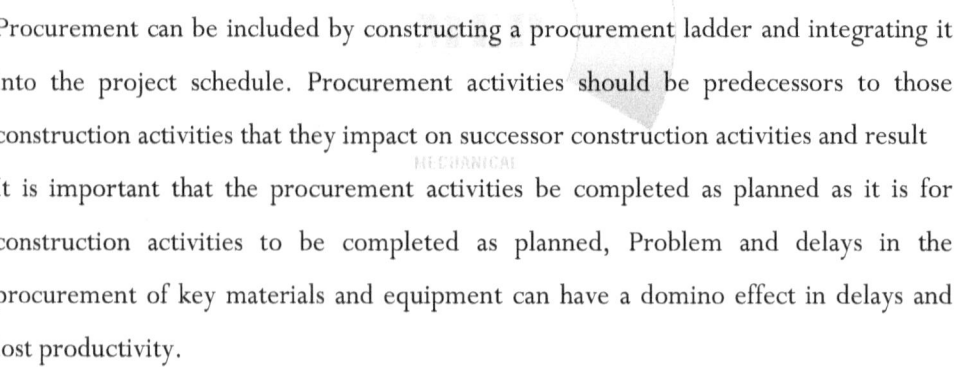

Procurement can be included by constructing a procurement ladder and integrating it into the project schedule. Procurement activities should be predecessors to those construction activities that they impact on successor construction activities and result

It is important that the procurement activities be completed as planned as it is for construction activities to be completed as planned, Problem and delays in the procurement of key materials and equipment can have a domino effect in delays and lost productivity.

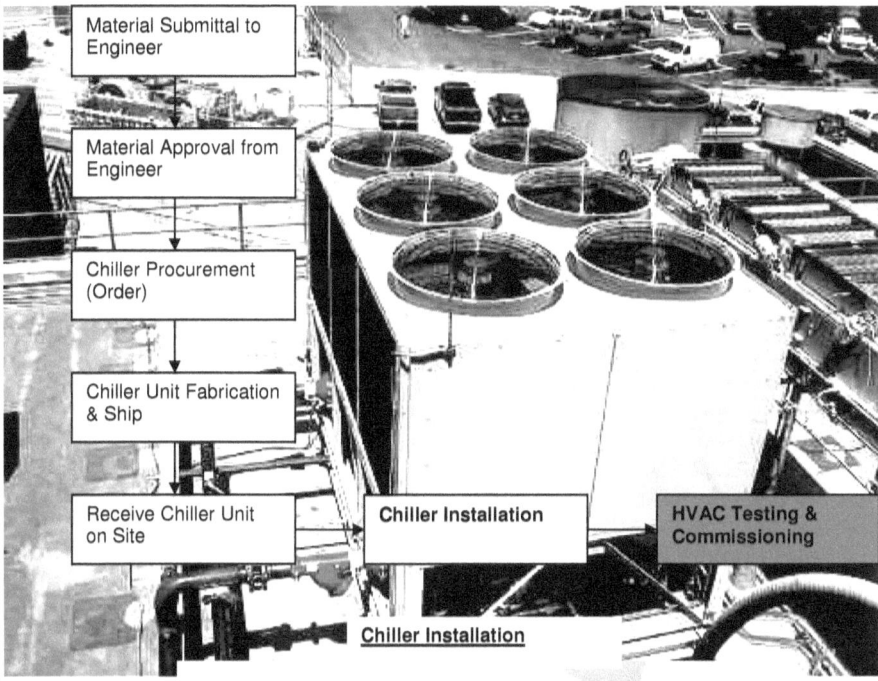

A simple example of chiller unit procurement in figure above illustrates the delay in chiller unit procurement & delivery can have a domino effect in delays of the chiller installation and HVAC Testing & Commissioning.

This section covers the MEP material procurement & delivery process for the material which is critical to the success of the construction project. The duration mentioned in this section is the minimum duration required for material procurement.

| S.No. | Description | Unit | Duration Requirement - 1st Delivery On Site | | | | |
|---|---|---|---|---|---|---|---|
| | | | Submittal | Approval | Procurement | Manufacturing & Delivery | Total |
| | **PIPES** | | | | | | |
| PIP-01 | Pipe for Fire Fighting System | mtr | 21 | 35 | 14 | 60 | 130 |
| PIP-02 | Pipe for Sanitary Water System | mtr | 14 | 14 | 9 | 30 | 67 |
| PIP-03 | Pipe for Rain Water System | mtr | 14 | 14 | 9 | 30 | 67 |
| PIP-04 | Pipe for Cold Water Supply System | mtr | 18 | 14 | 12 | 30 | 74 |
| PIP-05 | Pipe for Hot Water Supply System | mtr | 18 | 14 | 12 | 30 | 74 |
| PIP-06 | Pipes for Chiled Water Supply / Return | mtr | 21 | 35 | 14 | 60 | 130 |
| PIP-07 | Pipes for Natural Gas | mtr | 14 | 21 | 12 | 45 | 92 |
| PIP-08 | Pipes for LPG Gas | mtr | 14 | 21 | 12 | 45 | 92 |
| | **VALVES** | | | | | | |
| VAL-01 | Double Regulation Valves | nos | 21 | 35 | 14 | 60 | 130 |
| VAL-02 | Valves for Fire Fighting System | nos | 21 | 21 | 12 | 60 | 114 |
| VAL-03 | Pressure Reducing Valve Station | nos | 21 | 21 | 14 | 60 | 116 |
| VAL-04 | Automatic Air Vent | nos | 14 | 14 | 12 | 30 | 70 |
| VAL-05 | Valves for Sanitary Water System | nos | 14 | 14 | 12 | 30 | 70 |
| VAL-06 | Valves for Rain Water System | nos | 14 | 14 | 12 | 30 | 70 |
| VAL-07 | Valves for Water Supply System | nos | 14 | 14 | 12 | 30 | 70 |
| VAL-08 | Valves for Chiled Water Supply / Return | nos | 21 | 30 | 14 | 60 | 125 |
| VAL-09 | Landing Valves | nos | 14 | 14 | 14 | 45 | 87 |
| VAL-10 | Valves for Natural / LPG Gas | nos | 14 | 14 | 14 | 45 | 87 |
| | **PUMPS** | | | | | | |
| PUM-01 | Sump Pump | nos | 21 | 14 | 14 | 60 | 109 |
| PUM-02 | Water Distribution Pumps | nos | 21 | 14 | 14 | 90 | 139 |
| PUM-03 | Sewage Pumps | nos | 21 | 14 | 14 | 90 | 139 |
| PUM-04 | Booster Pumps | nos | 21 | 14 | 14 | 90 | 139 |
| PUM-05 | Fire Water Pumping Station | nos | 30 | 21 | 14 | 120 | 185 |
| PUM-06 | Chilled Water Pumps | nos | 30 | 35 | 14 | 120 | 199 |
| | **MAJOR MEP EUIPMENTS** | | | | | | |
| EQP-01 | Fan Coil Units | nos | 21 | 35 | 14 | 90 | 160 |
| EQP-02 | Air Handling Units | nos | 21 | 35 | 14 | 120 | 190 |
| EQP-03 | Chillers | nos | 21 | 45 | 21 | 120 | 207 |
| EQP-04 | Water Heat Exchanger | nos | 21 | 35 | 14 | 90 | 160 |
| EQP-05 | Air to Air Heat Exchanger | nos | 21 | 35 | 14 | 90 | 160 |
| EQP-06 | Air Scrubbers | nos | 14 | 21 | 12 | 60 | 107 |
| EQP-07 | Expansion Tanks | nos | 14 | 21 | 12 | 60 | 107 |
| EQP-08 | Boiler / Burner | nos | 14 | 21 | 12 | 90 | 137 |
| EQP-09 | Calorifier | nos | 14 | 35 | 12 | 90 | 151 |
| EQP-10 | Generator | nos | 21 | 45 | 14 | 120 | 200 |
| EQP-11 | Motor Control Center (MCC) | nos | 21 | 35 | 14 | 120 | 190 |
| | **LPG / NATURALGAS ITEMS** | | | | | | |
| GAS-01 | Gas Storage tank | nos | 14 | 35 | 14 | 75 | 138 |
| GAS-01 | Gas Leak Detect System | nos | 14 | 35 | 14 | 75 | 138 |

| S.No. | Description | Unit | Duration Requirement - 1st Delivery On Site | | | | |
|---|---|---|---|---|---|---|---|
| | | | Submittal | Approval | Procurement | Manufacturing & Delivery | Total |
| | **FIRE FIGHTING & PLUMBING ITEMS** | | | | | | |
| FFP-01 | Sprinkler | nos | 14 | 21 | 14 | 60 | 109 |
| FFP-02 | Zone Fire Alarm Flow Switch | nos | 14 | 21 | 14 | 60 | 109 |
| FFP-03 | Fire Hose Cabinet | nos | 14 | 21 | 14 | 75 | 124 |
| FFP-04 | Fire Hose Reel | nos | 14 | 21 | 14 | 75 | 124 |
| FFP-05 | Portable Fire Extinguisher | nos | 21 | 35 | 14 | 60 | 130 |
| FFP-06 | Dry Chemical (ABC) System | nos | 21 | 35 | 14 | 60 | 130 |
| FFP-07 | FM-200 Extinguishing System | nos | 21 | 35 | 14 | 90 | 160 |
| FFP-08 | Water Meter | nos | 14 | 21 | 14 | 60 | 109 |
| FFP-09 | Irrigation System | nos | 21 | 45 | 14 | 90 | 170 |
| FFP-10 | GRP Water Tanks | nos | 14 | 21 | 14 | 60 | 109 |
| FFP-11 | Floor Drain | nos | 14 | 14 | 14 | 45 | 87 |
| FFP-12 | Sanitary Fixtures | nos | 14 | 35 | 14 | 60 | 123 |
| FFP-13 | Pipe Support & Clamp | nos | 12 | 21 | 14 | 30 | 77 |
| FFP-14 | Grease Trap | nos | 14 | 21 | 14 | 45 | 94 |
| FFP-15 | Water Hammer Arrestor | nos | 14 | 21 | 14 | 45 | 94 |
| FFP-16 | Electric Water Heaters | Nos | 14 | 35 | 14 | 60 | 123 |
| | **HVAC ITEMS** | | | | | | |
| VAC-01 | Ducts | sq.m | 14 | 35 | 14 | 45 | 108 |
| VAC-02 | Flexible Duct | m | 14 | 35 | 14 | 45 | 108 |
| VAC-03 | Duct Insulation | sq.m | 14 | 14 | 14 | 45 | 87 |
| VAC-04 | Fire Damper | nos | 14 | 14 | 14 | 60 | 102 |
| VAC-05 | Motorized Smoke Damper | nos | 14 | 35 | 14 | 60 | 123 |
| VAC-06 | Volume Control Damper | nos | 14 | 14 | 14 | 60 | 102 |
| VAC-07 | Dack Draft Damper | nos | 14 | 14 | 14 | 60 | 102 |
| VAC-08 | Gravity Damper | nos | 14 | 14 | 14 | 60 | 102 |
| VAC-09 | Fire Damper | nos | 14 | 35 | 14 | 60 | 123 |
| VAC-10 | Sound Attenuator | nos | 14 | 35 | 14 | 60 | 123 |
| VAC-11 | Accoustic Liner for Duct | sq.m | 14 | 14 | 14 | 45 | 87 |
| VAC-12 | Access Doors | nos | 14 | 14 | 14 | 45 | 87 |
| VAC-13 | Air Filters | nos | 21 | 14 | 14 | 60 | 109 |
| VAC-14 | Vibration Isolators | nos | 14 | 14 | 14 | 60 | 102 |
| VAC-15 | Control Valve & Thermostst | nos | 14 | 21 | 14 | 60 | 109 |
| VAC-16 | Centrifugal Fans | nos | 14 | 35 | 14 | 60 | 123 |
| VAC-17 | Centrifugal In LineFans | nos | 14 | 35 | 14 | 60 | 123 |
| VAC-18 | Foor Top Fans | nos | 14 | 21 | 14 | 60 | 109 |
| VAC-19 | Portable Air Fans | nos | 14 | 21 | 14 | 60 | 109 |
| VAC-20 | Axial Fans | nos | 14 | 21 | 14 | 60 | 109 |
| VAC-21 | Staircase Pressure Fan Installation | nos | 14 | 35 | 14 | 60 | 123 |
| VAC-22 | Diffuser | nos | 14 | 14 | 14 | 45 | 87 |
| VAC-23 | Grilles & Registers | nos | 14 | 14 | 14 | 45 | 87 |
| VAC-24 | Building Management System | lot | 30 | 45 | 21 | 90 | 186 |

ARAB MEP
www.arabmep.com

| S.No. | Description | Unit | Duration Requirement - 1st Delivery On Site | | | | |
| --- | --- | --- | --- | --- | --- | --- | --- |
| | | | Submittal | Approval | Procurement | Manufacturing & Delivery | Total |
| | **ELECTRICAL ITEMS** | | | | | | 0 |
| ELE-01 | ATS Panel | nos | 21 | 35 | 14 | 60 | 130 |
| ELE-02 | Main Panel Boards MDBs | nos | 21 | 35 | 14 | 60 | 130 |
| ELE-03 | Capacitor Bank | nos | 21 | 35 | 14 | 60 | 130 |
| ELE-04 | Intermediate Distribution Boards SMDBs | nos | 21 | 35 | 14 | 60 | 130 |
| ELE-05 | Distribution Panel Boards DBs | nos | 14 | 21 | 14 | 45 | 94 |
| ELE-06 | UPS Distribution Boards UDBs | nos | 14 | 35 | 14 | 60 | 123 |
| ELE-07 | Earthing and Lighting Protection | item | 14 | 35 | 14 | 45 | 108 |
| ELE-08 | Power Cables | mtr | 14 | 21 | 14 | 45 | 94 |
| ELE-09 | Cable Trays | mtr | 14 | 14 | 14 | 30 | 72 |
| ELE-10 | Cable Trunk | mtr | 14 | 14 | 14 | 30 | 72 |
| ELE-11 | Wiring Devices | item | 14 | 35 | 14 | 45 | 108 |
| ELE-12 | Conduits | mtr | 14 | 14 | 14 | 30 | 72 |
| ELE-13 | Light Fixtures | nos | 14 | 35 | 14 | 60 | 123 |
| ELE-14 | Emergency Lighting | nos | 14 | 35 | 14 | 45 | 108 |
| ELE-15 | Self Contained Emergency Lighting | nos | 14 | 35 | 14 | 45 | 108 |
| ELE-16 | Disconnect Switch and Cables | nos | 14 | 21 | 14 | 30 | 79 |
| ELE-17 | Telecommunication System | nos | 21 | 35 | 14 | 60 | 130 |
| ELE-18 | E-PABX System | lot | 21 | 35 | 14 | 60 | 130 |
| ELE-19 | External Lighting | lot | 14 | 21 | 14 | 60 | 109 |
| ELE-20 | Dimming Panel | nos | 21 | 35 | 14 | 60 | 130 |
| ELE-21 | Fire Alarm | nos | 14 | 21 | 14 | 45 | 94 |
| ELE-22 | CCTV System | lot | 21 | 35 | 14 | 60 | 130 |
| ELE-23 | Access Control System | lot | 21 | 35 | 14 | 60 | 130 |
| ELE-24 | Centeral Battery | nos | 21 | 21 | 14 | 45 | 101 |
| ELE-25 | Bus Duct | ml | 21 | 35 | 14 | 45 | 115 |
| ELE-26 | Cable & Wiring Identification System | lot | 14 | 14 | 14 | 30 | 72 |

# SECTION 7

## DEWA REGULATION

Procedure to obtain DEWA's approval for Electricity LV Wiring Installation

1. Take the prescribed application form for drawing approval.
2. Complete the application form and submit for approval, the drawings in the standard formats specified in clause 1.7 of DEWA Regulations for Electrical Installation (attached with this manual) supported by the following:-
   a) Copy of Building Permit "NOC"
   b) Site setting out layout/ key plan
   c) Connected Load & Maximum demand Schedules
   d) Single Line Diagram
   e) Dimensional layout of electrical switch rooms/ metering enclosures
   f) Wiring layouts
   g) Load distribution schedules
   h) Plan of substation location (if requirement of substation is indicated in DEWA's Building Permit "NOC".
   i) Other drawings/ technical specifications, as applicable
3. Collect the drawings duly approved or commented from DEWA.
4. Applications returned requiring compliance should be resubmitted, duly complied with as per revised delivery.

New Connection Registration

1. Obtain the prescribed application for supply of Electrical and Water, along with the checklist of the documents to be submitted.
2. Submit the following documents along with the completed application form for registration:
   a. Dubai Municipality approved Affection Plan
   b. Copy of identification documents: trade license
   c. Authorization letter for representatives
   d. DEWA's approved connected load/ maximum demand schedule
   e. DEWA's valid Building permit NOC (within six months from the date of issue) with:

| | |
|---|---|
| PLANNING DEPARTMENT | |
| Rev.: 0 | **MEP PLANNING MANUAL** |
| Date of Issue: Feb 2007 | |
| Doc No.: MEP-01-07 | |

      i. Copy of DEWA's approval plan of substation location (if requirement of substation is indicated in NOC)

      ii. Copy of DEWA approval site setting out layout/ plan showing location of meter box/electrical.

   f. Submit in writing from landlord/ consultant. Electrical contractor the following:

      i. Power and/or water requirement date / project completion date

      ii. Name and position of coordinate Engineer/ Representative

3. Collect the registration slip, which indicates the date when the estimate will be ready.

# Note: - <u>DEWA Regulations for Electrical Installation attached</u>

# SECTION 8

# DUBAI CIVIL DEFENCE REGULATION

## Introduction

- Inspects all installations to ensure compliance with prevention and safety conditions according to laws, decisions and directives issued in this regard.

- Conducts regular and emergency inspection to ensure compliance with prevention and safety conditions, and prepares inspection reports including all data on conditions especially alarm and fire extinguishing systems, electricity, staircases, outlets, pumps and guiding signboards.

- Reviews licensing documents of establishments which sell and deal in civil defense equipment, and verifies weights, types and integrity of equipment.

- Supervises execution of civil defense plans for civilian protection.

- Supervises and follows up alarm systems and their methods with the purpose of developing them, as well as test them to ensure their operativeness according to stated timings.

- Articipates in determining technical specifications and standards for public and private shelters.

- Participates with relevant authorities in preparation of estimates for strategic reserves of foodstuffs and living provisions in normal and emergency cases as well as in disasters.

- Arranges preparations for communications and coordination with parties involved in civil defense measures, including coordination with hospitals, emergency medical aid centers and social services centers.

- Organizes inspection operations and procedures starting form delivering of licensing applications, classifying, numbering and registering them, examines applications especially license copies, site plans and certificates of equipment and refers them for inspection, as well as prepares inspection reports and issues required letters for the purpose of inspection and to installations owners according to a written request indicating fulfillment of prevention and safety conditions.

- Arranges and maintains transactions filing according to a classification indicating numbers and types of installations such as hotels, institutes, companies, gas distribution establishments, fuel supply stations, factories and commercial and professional establishments.

- Keeps and continuously reviews a special undertakings register in order to ensure their implementation in normal conditions and after accidents.

- Registers layouts of building, decoration works and certificates of completion.

- Observes principles for dealing with applicants and representatives of contracting companies.

## Requirements for Decoration Approval

### First: Decoration Plan

Elevation plan:

Including partitions and modifications + decorations and the activity of the place.

## Sections details for:

- Ceiling.

- Partitions.

- False ceiling components (if any).

Note : Mention the materials used.

Second: Ceiling Plan:

Indicate the materials used in covering the ceiling + fire system and fire fighting which existing there.

## Third: A Letter from approved decoration office, mentioning the following:

The fixed period to finish decoration works.

Building location and owner's name.

Definition to decoration works that will be carried out.

Fourth: Submit Three Copies of Design Each Copy Includes:

Elevation plan.

Ceiling plan.

Note: Transaction should be submitted to Directorate - Engineering Section,

Engineering Section, Dubai Civil Defence.

## Fifth: Contact Dubai Civil Defence Inspection section, after finishing decoration work. To have an appointment for inspection, and you should attached :

Approved decoration design.

A letter from decoration office.

Trade License.

# Approval of drawings

- Submission of project's drawings by a specialized engineer to get them approved by civil defence.

- The drawings will be studied by safety engineer from civil defence. If the drawings meet safety requirements, they will be approved and stamped by civil defence. However, if the drawings fail to meet safety requirements, they will be returned for completion.

- After finishing decoration work, the client is given an appointment for inspection, which will be followed by approval if safety requirements are met.

# Licensing (approving safety equipment and fire doors)

1. Clients are requested to submit the required documents at the civil defence to approve companies dealing with or circulating safety equipment and fire doors:

   - Approved agent.

   - Approved factory.

   - Approved distributor.

   - Approved installation and maintenance technician.

   - Approved safety fire safety consultants.

2. Technical and administrative study will be conducted on the transaction's file.

If the file meets safety requirements, it will be sent to civil defence headquarters to issue the company with licence. In the event of failure the file will be returned to the client.

## Explosive and hazardous materials

1. Fire safety companies are requested to submit the following documents to civil defence:

- Vehicles registered to transport hazardous materials.

- Licence allowing import, export and transport of hazardous materials.

- Lincence allowing to deal/explode firecrackers and balloons

- Licence of hazardous materials storage/warehouse.

- Licence for reservoirs or containers to keep inflammable substances.

- Lincence for petrol stations and gas refilling factories.

- Licence allowing the company to circulate hazardous materials.

2. An appointment is set by the inspection department of the civil defence to check these licences. If safety conditions are met, clients are issued with approval documents or other concerned departments being notified about safety compliance.

## Comprehensive prevention

1. The approved fire safety companies should be present at at civil defence counters

to submit fire safety contracts for approval.

2. The safety officer will randomly check the contract to ensure safety measures, which are mentioned in it are correct and get it approved from the head of safety section. Following payment of approval fees, the fire safety company collects the contract.

## Completion of building construction and decoration

1. The consultant should present the main building's drawings, fire fighting and alarm systems implementation plan and other allied equipment to be approved by safety engineer at the municipality or civil defence department.

2. When the building is completed, the contractor or whoever deputises him should apply for an appointment with civil defence counters to inspect the building and check whether safety systems (fire fighting, alarm and maintenance certificates) in the transaction match those of the completed building. The contractor should supply civil defence with a sketch map that locates the building.

3. If safety conditions are met, the civil defence licenses the building and officially recommends the municipality for quick completion of the transaction. If safety conditions are not met, however, the civil defence reschedules inspection appointment.

## Law enforcement

1. If the client fails to complete safety measures at the building following the safety

| | |
|---|---|
| PLANNING DEPARTMENT | |
| Rev.: 0 | |
| Date of Issue: Feb 2007 | **MEP PLANNING MANUAL** |
| Doc No.: MEP-01-07 | |

officer's inspection, the client will be advised and given designated period of time to rectify. If the client fails again, he will be issued with warning.

2. The client will be summoned to check whether safety measures are met or not. If the client meets the safety measures, the civil defence will issue him with a letter to the Department of Tourism and Commerce Marketing (DTCM) for approval to renew the contract. However, if the client fails to meet safety requirements a letter will be directed to the DTCM to decline the renewal or the transaction file will be referred to the public prosecution.

**Note:- For more information and details of Dubai Civil Defence please visit their web site**  http://www.dcd.gov.ae

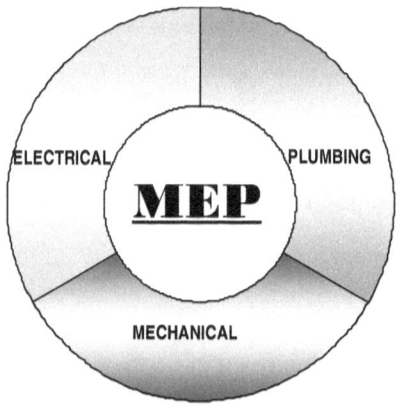

# MEP PLANNING MANUAL

## (PART – II)

## *A Guide to the Project Planning & Installation related to MEP Works*

*First Edition (Feb, 2007)*

| | |
|---|---|
| PLANNING DEPARTMENT | |
| Rev.: 0 | **MEP PLANNING MANUAL** |
| Date of Issue: Feb 2007 | |
| Doc No.: MEP-01-07 | |

# PREFACE OF THE FIRST EDITION

This overall methodology for Mechanical, Electrical & Plumbing (MEP) installation in buildings, towers & villa's are include in method statement. This method statement is intended to reflect our understanding of the requirements of the project and the general construction sequences that we will follow. It will provide information of reference documents and equipment required in installation stage.

This MEP method statement has been prepared by HEE professionals and got the approval from the Engineers & the Clients in our previously executed projects. With the help of this method statement it's easy to plan the activity in more professional, technical and smooth manner to maintain the quality of work, safety requirements implementation and time completion.

Each method statement is given unique code number for easy identification i.e.

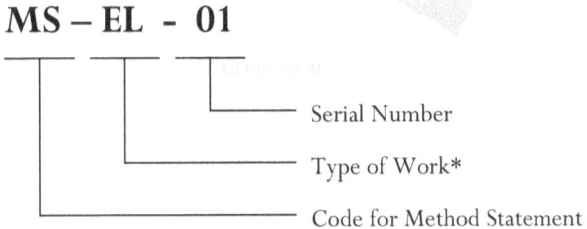

MS – EL – 01

- Serial Number
- Type of Work*
- Code for Method Statement

\* EL  : Electrical

AC  : Mechanical – HVAC

PL  : Plumbing

FF  : Fire Fighting

| PLANNING DEPARTMENT | |
|---|---|
| Rev.: 0 | |
| Date of Issue: Feb 2007 | **MEP PLANNING MANUAL** |
| Doc No.: MEP-01-07 | |

# <u>INTRODUCTION</u>

This Part-II is divided into four sections:-

Section 9   :      Method Statement for Electrical Installation

Section 10  :      Method Statement for Mechanical (HVAC) Installation

Section 11  :      Method Statement for Plumbing Installation

Section 12  :      Method Statement for Fire Fighting Installation

www.arabmep.com

# Section 9

# Method Statement for Electrical Installation

## Code No. : MS-EL-01

# Method Statement
# Installation of PVC Conduits Concealed in
# Concrete Slab and Columns

# METHOD STATEMENT FOR INSTALLATION OF PVC CONDUITS IN CONCRETE SLAB & CONCRETE WALLS

## SCOPE AND PURPOSE

This method statement covers the site installation of the PVC Conduits in floor slab & concrete wall and the requirements of the checks to be carried out. This procedure defines the method used to ensure that all conduit and associated accessories: bends, tees, couplers, reducers and all accessories associated with systems installed are correct and acceptable.

## REFERENCE DOCUMENTS

- Approved Shop Drawings
- Approves Material Submittals

## ASSOCIATED WORKS

- Power & Lighting Wiring

## RESPONSIBLE PERSONNEL

- Construction Manager
- Project Electrical Engineer
- Site Electrical Engineer
- Site Supervisor
- Technicians Qualified in the trade

## METHOD OF PRE INSTALLATION

- Ensure that approved material required to carryout work will be available.
- Prior to commencement of work, area and access will be inspected to confirm that the site is ready to commence the work
- All relevant documentation and material applicable to particular section of works will be checked by site engineer before commencement.
- Physical verification of material will be carried out for any damages prior to taking from stores.
- The site engineer / supervisor will give necessary instruction to tradesman and provide necessary construction / shop drawings.
- The site engineer / supervisor will also check that tools and equipments available are in compliance to contract requirements.
- The site supervisor also explains tradesman regarding safety pre-cautions to be observed.

## METHOD OF INSTALLATION

1. Supervisor will ensure that all the grid lines and datum lines are marked by the surveyor and all the route and marking are based on the datum line provided by the surveyor.
2. Supervisor will carryout a site survey and marks the route of conduits as per approved drawings. In the event that there are any discrepancies or difficulties in executing the work, these will be drought to the notice of engineer for corrective action.
3. PVC conduits of correct sizes will be used for concealment of electrical services in the concrete walls & slabs.
4. Cast in conduits are of high impact and of approved make and conform to BS 4607 Part 1, 3 & 5, BS 6099 Parts 1 & 2 and BS 5490.
5. The minimum conduit size used will be 20 mm dia. The conduit size selection is completely as per technical specifications, local statutory authority regulations and as per approved shop drawings.
6. Wherever conduit to be installed in vertical walls the run of conduit will be kept straight,
7. The bending of conduits will be done using proper bending springs and the straight lengths of conduits are coupled using standard couplers glued using good quality PVC sealant.
8. All the cast in conduits will be firmly bended to the steel rebar' and conduits is placed sandwiched between rebars so that future drilling of anchors should not damage cast in conduits and the installed wires.
9. Near DB end conduits are terminated on to thick thermocol placed under concrete so that future alignment of conduit will be easier.
10. Wherever cast in conduits to be terminated on to back boxes standard conduit accessories approved in the material submittal will be used.
11. All the due care will be taken to ensure non blocking of cast in conduits during concreting and contractor's representative will be present at the throughout of concrete period to ensure no conduits gets damaged during casting of concrete slabs.
12. The guide line for conduit size selection will be as per the following table.

| Cross Sectional Area of Conductors | Size of Conduit | | |
| --- | --- | --- | --- |
| | Maximum number of cable drawn | | |
| | 20 | 25 | 32 |
| 1.5 mm2 | 7 | 12 | - |
| 2.5 mm2 | 5 | 9 | 12 |
| 4.0 mm2 | 3 | 6 | 9 |
| 6.0 mm2 | - | 5 | 8 |

## QUALITY

- QCE along with Project Engineer and Site Supervisor will monitor that all components are installed as per contract specifications and approved submittals.

- Inspection Request shall be submitted to Consultant after completion of installation of cable tray before pulling of any wires

## SAFETY

- Work will commence as per safety regulations laid down in the contract specification and project safety plan.
- Proper safety harness to be used and secured, if required.
- All personal protective equipment shall be used as appropriate according to the nature of job.
- Housekeeping shall be of good standard and all cut lengths and debris shall be removed.

www.arabmep.com

## Code No. : MS-EL-02

# Method Statement
# G.I & PVC Conduit on Surface

# METHOD STATEMENT FOR G.I. & PVC SURFACE CONDUITING

## SCOPE AND PURPOSE

This method statement covers the site installation of the G.I. & PVC Surface Conduiting for small power and lighting and requirements of checks to be carrier out.

## REFERENCE DOCUMENTS

- Approved Shop Drawings
- Approves Material Submittals
- Approved method statement for wire pulling and earthing.

## APPLICABLE LOCATION

- Surface GI Conduiting is generally installed in Electrical rooms and plant rooms, lift motor rooms, sub-stations and outside buildings, partially open ceilings, service areas.
- PVC Conduits are used only the area covered by false ceiling.
- Installation of GI surface conduits shall be done in accordance with the project specifications and drawings.

## MATERIALS

- GI conduit and accessories – galvanized conforming to BS 4568 part 1 class 4.
- Metal boxes enclosing electrical accessories confirms to BS 4662.

## EQUIPMENT / TOOLS

- Measuring tape
- Drilling machine
- Pipe bending machine
- Threading machine with die-set
- Helical bend springs
- Hammer & Hack saw

## RESPONSIBLE PERSONNEL

- QA / QC Engineer
- Project Electrical Engineer
- Site Electrical Engineer
- Site Supervisor

- Electrician
- Helpers

## METHOD OF STORAGE

- Conduit shall be lowered not dropped to the ground.
- Conduit shall be given proper support at all times and shall be stacked on flat surface. Manufacturer's instructions shall be followed as applicable.
- Timber support shall be placed at an interval of one meter.
- Conduits shall be protected from direct sunlight and moisture to avoid deterioration.
- Conduits shall be stored with proper end caps.
- Conduits shall be stacked size wise.

## METHOD OF PRE INSTALLATION

- Ensure that approved material required to carryout work will be available.
- Proper materials including GI conduits and accessories are with drawn from stores according to approved shop drawing and good engineering practices.
- Physical verification of materials will be carried out for any damages prior to taking from stores and also prior to installation.
- Prior to commencement of work, area and access will be inspected to confirm that site is ready to commence the work.
- All relevant documentation and material applicable to particular section of will be checked by site supervisor prior to commencement of work.
- The site engineer/ site supervisor will give necessary instruction to Electrician and provide necessary approved construction/ shop drawings.
- The site engineer/ site supervisor will also check that proper tools and equipment are available to carry out the work and are in compliance with contract specification
- The site supervisor also explains the electrician regarding safety precautions to be observed.

## METHOD OF INSTALLATION

1. Supervisor will carryout a site survey and marks the route of conduit as per approved drawing. In the event that there are any discrepancies of difficulties in executing the work, these will be brought to the notice of project engineer for corrective action.
2. Standard length of conduits shall be cut to the required length.
3. For G.I. conduits threaded shall be done using a threaded machine and correct size die-set. Threaded will be kept to a minimum when showing from coupling and boxes.
4. Cold galvanizing paint is applied to the threaded part of G.I. conduit just before fixing.

| PLANNING DEPARTMENT | |
| --- | --- |
| Rev.: 0 | **MEP PLANNING MANUAL** |
| Date of Issue: Feb 2007 | |
| Doc No.: MEP-01-07 | |

5. Where required conduits of size 20mm and 25mm shall be bent to the required radius using manual bending machine. Manufactured bends are used for conduits size 32mm and 50mm.

6. Conduits are fixed to the building fabric by means of distance bars saddle with appropriate metal screws and plugs. A space of 3mm minimum shall be maintained between conduit and the building surface.

7. Check all mechanical connections are internally smooth for pulling wiring in no burrs or sharp edge will be allowed.

8. The route of this steel conduits are restricted to horizontal and vertical runs except for the areas where approved to follow the line of an architecture.

9. Conduit support are fixed in regular interval as mentioned in table:-

### Spacing of Conduit Supports

| Conduit Size (mm) | Maximum Distance Between Support (M) | | | |
| --- | --- | --- | --- | --- |
| | Rigid Steel | | Pliable | |
| | Horizontal | Vertical | Horizontal | Vertical |
| 20 | 1.75 | 2.0 | 0.4 | 0.6 |
| 25 | 1.75 | 2.0 | 0.4 | 0.6 |
| 32 | 2.0 | 2.25 | 0.6 | 0.8 |

10. Ensure the conduit installations are in straight line.

11. Conduits shall be fixed so that no water enters, if it is not practicable a 3mm hole is drilled at the lowest point of the conduit to drain the water out.

12. All terminal boxes are marked on the appropriate location (i.e. wall or ceiling) as per approved shop drawing and fixed with metal screws and plugs. Suitable braze bushes are used where conduit enters the boxes to avoid any damage to the wires.

13. Wherever necessary draw wires shall be pulled into conduit runs and kept at pull boxes for future use.

14. Any rusting of steel conduits occurred during construction shall be removed.

## QUALITY

- QCE along with Project Engineer and Site Supervisor will monitor that all components are installed as per contract specifications and approved submittals.
- Inspection Request shall be submitted to Consultant after completion of installation of cable tray before pulling of any wires

## SAFETY

- Work will commence as per safety regulations laid down in the contract specification and project safety plan.
- Proper safety harness to be used and secured, if required.
- All personal protective equipment shall be used as appropriate according to the nature of job.

## Code No. : MS-EL-03

# Method Statement
# Cable tray & Ladders

| PLANNING DEPARTMENT | |
|---|---|
| Rev.: 0 | |
| Date of Issue: Feb 2007 | **MEP PLANNING MANUAL** |
| Doc No.: MEP-01-07 | |

# METHOD STATEMENT FOR CABLE TRAY & LADDERS

## SCOPE AND PURPOSE

This method statement covers the site installation of the cable tray & ladders and requirements of checks to be carrier out.

## REFERENCE DOCUMENTS

- Approved Shop Drawings
- Approves Material Submittals
- Approved method statement for cable laying and earthing

## APPLICABLE LOCATION

- Cable Tray system is generally installed in electrical rooms, plant rooms, service corridors as detailed in approved shop drawings.

## MATERIALS

- Cable Trays and associated fittings (Hot dip galvanized conforming to BS 1449 part 1 BS 729)
- Support system (Bracket work, Support Rod, Anchors etc.)

## EQUIPMENT / TOOLS

- Drilling machine
- Grinding machine
- Electrician hand tools

## RESPONSIBLE PERSONNEL

- Project Electrical Engineer
- Site Electrical Engineer
- QA / QC Engineer
- Site Supervisor
- Electrician & Helpers

## METHOD OF STORAGE

- All material received at site shall be inspected and ensure that the material are as per approved material submittal.

- Any discrepancies, damages etc. Shall be notified and reported for further action.
- Materials found not suitable for the project are removed from site immediately.
- Cable tray and ladders are store in horizontal position on flat surface with timber support placed at an interval of one meter and covered to protect from moisture and direct sunlight.

## METHOD OF PRE INSTALLATION

- Ensure that approved material required to carryout work will be available.
- Proper materials including cable trays, ladders and associated fittings accessories are with drawn from stores according to approved shop drawing and good engineering practices.
- Physical verification of materials will be carried out for any damages prior to taking from stores and also prior to installation.
- Prior to commencement of work, area and access will be inspected to confirm that site is ready to commence the work.
- All relevant documentation and material applicable to particular section of will be checked by site supervisor prior to commencement of work.
- The site engineer/ site supervisor will give necessary instruction to Electrician and provide necessary approved construction/ shop drawings.
- The site engineer/ site supervisor will also check that proper tools and equipment are available to carry out the work and are in compliance with contract specification
- The site supervisor also explains the electrician regarding safety precautions to be observed.

## METHOD OF INSTALLATION

15. After the civil clearance to proceed with MEP installations, ensure the area is clean and ready to start the works.
16. Mark the trays and ladders routes as per approved shop drawing; ensure these are of horizontal & vertical runs only.
17. Co-ordinate the routes, levels and ensure there are no clashes with other services.
18. Maintain enough clearance for cable pulling and any access for future maintenance.
19. Mark the support, fix the threaded rod supports with appropriate metal plugs, and then fix the 'L' angles / Slotted 'C' channels with nuts. A maximum of 1.2 M distance is maintained between the supports to avoid sagging of trays and ladders. Provide adequate supports for bends, branches and offsets.
20. Cut the standard length / ladder to required length with appropriate cutting tools. Use mushroom head screws on the cable route to avoid the cable insulation damage during pulling.
21. All the fittings shall be from manufacturer.
22. Expansion provision shall be provided at all the building expansion joints.
23. Manufacturer's instruction for installation shall be followed.
24. Approved fire sealant shall be provided wherever the installation crosses the fire rated walls.
25. Sleeves shall be provided at all the wall crossing.
26. Copper earth link shall be installed at every joint to maintain continuity throughout the installation.
27. Ensure the installation of tray / ladders are neat, in straight line. Trim the extra projected supports.

28. All sharp edges and burrs shall be cleaned for pulling the cables.
29. Treat the cut & Drilled part with zinc paint rich paint immediately after cutting and before installation.
30. Provide identification labels as specified to identify the service.
31. Inspection shall be offered fro QC verification.
32. The entire installed cable tray to be protected from damages.

## QUALITY

- QCE along with Project Engineer and Site Supervisor will monitor that all components are installed as per contract specifications and approved submittals.
- Inspection Request shall be submitted to Consultant after completion of installation of cable tray before pulling of any wires

## SAFETY

- Work will commence as per safety regulations laid down in the contract specification and project safety plan.
- Proper safety harness to be used and secured, if required.
- All personal protective equipment shall be used as appropriate according to the nature of job.

www.arabmep.com

## Code No. : MS-EL-04

# Method Statement
# Cable Trunking System

www.arabmep.com

| PLANNING DEPARTMENT | |
| --- | --- |
| Rev.: 0 | **MEP PLANNING MANUAL** |
| Date of Issue: Feb 2007 | |
| Doc No.: MEP-01-07 | |

# METHOD STATEMENT FOR CABLE TRUNKING SYSTEM

## SCOPE AND PURPOSE

This method statement covers the site installation of the cable trunking system and requirements of checks to be carrier out. This procedure defines the method used to ensure that all cable trunking and associated fittings: bends, tees, couplers, reducers, flanges, relevant bracket work, supports rods, anchors and all accessories associated with integrated cable management systems installed are correct and acceptable.

## REFERENCE DOCUMENTS

- Approved Shop Drawings
- Approves Material Submittals

## EQUIPMENT / TOOLS

- Measuring tape
- Drilling machine
- Grinding machine
- Electrician hand tools

## RESPONSIBLE PERSONNEL

- Project Electrical Engineer
- Site Electrical Engineer
- QA / QC Engineer
- Site Supervisor
- Electrician
- Helpers

## METHOD OF PRE INSTALLATION

- Ensure that approved material required to carryout work will be available.
- Proper materials including cable trays, ladders and associated fittings accessories are with drawn from stores according to approved shop drawing and good engineering practices.
- Physical verification of materials will be carried out for any damages prior to taking from stores and also prior to installation.
- Prior to commencement of work, area and access will be inspected to confirm that site is ready to commence the work.
- All relevant documentation and material applicable to particular section of will be checked by site supervisor prior to commencement of work.

- The site engineer/ site supervisor will give necessary instruction to Electrician and provide necessary approved construction/ shop drawings.
- The site engineer/ site supervisor will also check that proper tools and equipment are available to carry out the work and are in compliance with contract specification
- The site supervisor also explains the electrician regarding safety precautions to be observed.

## METHOD OF INSTALLATION

33. Work will be carried out as per manufacturer's recommendations.
34. Supervisor will carryout a site survey and marks the route of cable trunking as per approved drawings.
35. Dedicated cable trunking of size as approved in the shop drawing will be provided for lighting/ power and extra low voltage (ELV) services.
36. For connection of PVC/GI conduits to GI trunking installation details shown in approved shop drawing will be followed.
37. Cable trunking will be in accordance with BS 4678 Part 1 and made of galvanized steel.
38. Adjoining lengths of trunking will be correctly aligned and the two sides at right angles to the cover will be joined to the corresponding side of the trunking piece by means of an internal fish plate connector.
39. Standard manufacturer's fittings will only be used except when necessitated by site conditions, the consultant engineer's approval will be obtained.
40. All the joints of trunking both at straight lengths or accessories will be fitted with a copper bonding strap and will be secured by screw, nuts and washers. The earth bonding links will be external to trunking and will make good contact with the trunking and continuity will not depend on the contact through screws. Jointing screws will be installed with exposed thread and nut/spring washer external to the trunking.
41. Trunking fixing centers are not to exceed those listed below and only incase of 50x50 trunking there will be a center fixing only. All other sizes will have two fixing at appropriate fixing centers as given below:

| | Distance between Supports | |
| --- | --- | --- |
| Cross Sectional Area of Trunking | Steel Trunking | |
| | Horizontal | Vertical |
| 1. Upto 2500 mm2 | 1.20 M | 1.5 M |
| 2. Above 2500 mm2 and up to 6000 mm2 | 1.20 M | 1.8 M |
| 3. Above 6000 mm2 and up to 10000 mm2 | 2.3 M | 2.5 M |
| 4. More than 10000 mm2 | 3.0 M | 3.0 M |

42. All trunking accessory fittings will be secured not more than 150mm from the jointing points.
43. Additional support will be provided if required.

| PLANNING DEPARTMENT | |
| --- | --- |
| Rev.: 0 | **MEP PLANNING MANUAL** |
| Date of Issue: Feb 2007 | |
| Doc No.: MEP-01-07 | |

44. Cable tray will be inspected before pulling the wires.

45. Trunking will be installed in such a way that adequate clearance will be provided for access to wiring. Covers will be installed on to the exposed side of trunking. Where trunking installed horizontal plane with lid on to bottom side then conduit connection will be done from top and side.

46. Earth continuity will be available through out the trunking including bends and offsets.

47. Cable trunking crossing the fire rated walls will be provided with fire protection barriers of approved materials according to civil defense regulations and will be by main contractor's scope.

48. Any cuts made to the trunking will be debarred for rough surface and treated with cold galvanized paint.

49. The supervisor in charge and QC inspector will monitor the activities to ensure that all components indicated in the approved construction drawing installed as per the contract requirements and manufacture recommendations.

## QUALITY

- QCE along with Project Engineer and Site Supervisor will monitor that all components are installed as per contract specifications and approved submittals.
- Inspection Request shall be submitted to Consultant after completion of installation of cable tray before pulling of any wires

## SAFETY

- Work will commence as per safety regulations laid down in the contract specification and project safety plan.
- Proper safety harness to be used and secured, if required.
- All personal protective equipment shall be used as appropriate according to the nature of job.
- Housekeeping shall be of good standard and all cut lengths and debris shall be removed.

## Code No. : MS-EL-05

# Method Statement
# Installation of Earthing System

# METHOD STATEMENT FOR INSTALLATION OF EARTHING SYSTEM

## SCOPE AND PURPOSE

This method statement covers the site installation of the Earthing System

## REFERENCE DOCUMENTS

- Approved Shop Drawings
- Approves Material Submittals

## MATERIAL

- Earth rods, Earth rod clamps, Earth bars and Disconnecting Links

## RESPONSIBLE PERSONNEL

- Project Electrical Engineer
- Site Electrical Engineer
- Site Supervisor
- Technicians Qualified in the trade

## APPLICABLE LOCATION

- Basement, Ground level, Roof, Mechanical & Electrical rooms al all levels. All mechanical services exposed will be bonded to the earthing system.

## METHOD OF PRE INSTALLATION

- Ensure that approved material required to carryout work will be available.
- Prior to commencement of work, area and access will be inspected to confirm that the site is ready to commence the work
- All relevant documentation and material applicable to particular section of works will be checked by site engineer before commencement.
- Physical verification of material will be carried out for any damages prior to taking from stores.
- The site engineer / supervisor will give necessary instruction to tradesman and provide necessary construction / shop drawings.
- The site engineer / supervisor will also check that tools and equipments available are in compliance to contract requirements.
- The site supervisor also explains tradesman regarding safety pre-cautions to be observed.

| | |
|---|---|
| PLANNING DEPARTMENT | |
| Rev.: 0 | **MEP PLANNING MANUAL** |
| Date of Issue: Feb 2007 | |
| Doc No.: MEP-01-07 | |

## METHOD OF INSTALLATION

1. As per approved drawing identify location of earth pit.
2. Co-ordinate with civil contractor for installation of earth pits with approved fixing details.
3. The copper earth rod will be driven manually in earth. After achieving a minimum depth of approximate 3 meters, the earth resistance will be measured, If the earth resistance value is not satisfactory, the process of adding further earth electrodes shall be continued till expected resistance value of less than 1 ohm is achieved.
4. Adjacent earth electrodes shall be spread to atleast 1 length of one of the driven electrodes (6 mtrs) to achieve result of less that 1 ohm. Inspection request to be issued.
5. The earth pit shall be installed after completion of installation of earth rods and a clear gap of 50 mm shall be maintained between top of earth electrode and earth pit cover. The top of earth pit shall be in level with the finished floor level in the area.
6. The PVC sheathed single core earthing cables of specified sized as per shop drawing shall be laid between the earth pit and the earth bar inside the building and terminated with approved type lugs/ clamps.
7. The interconnection of earth pit shall be as per approved shop drawing.
8. All earthing connections shall be made after cleaning the surface thoroughly and tightness checks for each connection shall be performed.
9. Continuity of earth connections shall be checked for every link in the network by QA Engineer, Issue Inspection request.
10. The down-stream earthing connections from earth bars shall be made to the panel boards, frames and other equipment as per approved shop drawing.
11. Alongwith all power cables one earth cable of size as per approved shop drawing shall be laid and it shall be terminated to the waarth bat of the panel/ equipment which it feeds, in addition with local earthing from earth bar.
12. On completion of total earthing system and testing, inspection request will be submitted for approval to design consultant.

## INSTALLATION OF EQUIPOTENTIAL BONDING

1. The metallic frame of all electrical equipments shall be connected to the nearest earrh bar with a specific size of earth cable.
2. The earthing continuity of cable tray and trunking shall be maintained with earth links on each joints of cable trays and trunking shall be connected to earth bar with specified size of earth cable.
3. Flexible earth cable shall be used for the earthing connections when there is possibility of expansion/ contraction and also where vibrating equipment is installed.
4. The metallic water lines shall be bonded by an earthing cable of size not less than 6mm².
5. All bonding connections shall be checked for correct tightness and cleanliness.
6. Inspection Request will be issued for signature of consultants.

| | |
|---|---|
| PLANNING DEPARTMENT | |
| Rev.: 0 | **MEP PLANNING MANUAL** |
| Date of Issue: Feb 2007 | |
| Doc No.: MEP-01-07 | |

## QUALITY CONTROL

- In general, it shall be ensure by site engineer that product manufacturer's recommendations are followed and shall be monitored by QA/QC Engineer. However, the following points shall be ensured, in particular;
  - The appropriate and approved material is used
  - Skilled labor is used for application of the material.
  - Inspection request shall be raised for consultant's inspection.
  - QC inspection shall be carried out as per the installation checklist and manufacture's instructions.

## SAFETY

- Work will commence as per safety regulations laid down in the contract specification and project safety plan.
- Proper safety harness to be used and secured, if required.
- All personal protective equipment shall be used as appropriate according to the nature of job.
- Housekeeping shall be of good standard and all cut lengths and debris shall be removed.

## Code No. : MS-EL-06

# Method Statement
# Installation of Light Protection System

# METHOD STATEMENT FOR INSTALLATION OF LIGHTNING PROTECTION SYSTEM

## SCOPE AND PURPOSE

This method statement covers the site installation of the Lightning Protection System.

## REFERENCE DOCUMENTS

- Approved Shop Drawings
- Approves Material Submittals

## MATERIAL

- Material used for the system is as per BS 6651 : 1992

## RESPONSIBLE PERSONNEL

- Project Electrical Engineer
- Site Electrical Engineer
- Site Supervisor
- Technicians Qualified in the trade

## METHOD OF PRE INSTALLATION

- Ensure that approved material required to carryout work will be available.
- Prior to commencement of work, area and access will be inspected to confirm that the site is ready to commence the work
- All relevant documentation and material applicable to particular section of works will be checked by site engineer before commencement.
- Physical verification of material will be carried out for any damages prior to taking from stores.
- The site engineer / supervisor will give necessary instruction to tradesman and provide necessary construction / shop drawings.
- The site engineer / supervisor will also check that tools and equipments available are in compliance to contract requirements.
- The site supervisor also explains tradesman regarding safety pre-cautions to be observed.

## METHOD OF INSTALLATION

1. The dedicated rebar is selected from the pile cap and the cable is clamped with the rebar by CR730 clamp and connected to the dedicated rebar of the selected column as per approved shop drawing.

2. Fix G.I. Box 160x80x35 at 500mm from SSL (for testing purpose) on the column. The box shall be tied with reinforcement rod and the box cover will be flush with the finish wall.

3. The dedicated rebars of the column as per shop drawing shall be connected to a 70 sq.mm PVC cable using CR 705 Furse clamp for the extension till it reaches the roof.

4. Expanded polystyrene shall be applied to seal the hole within the concealed disconnect link box and tied with steel binding wire to prevent concrete / cement entering into the box.

5. Prior to concreting, earth continuity between reinforcement rods and dedicated rods shall be checked with a.d.c. ohm meter.

6. In each tower final test sheet shall be offered to consultant for witness and sign-off.

7. Once rebar has reached up to the height as shown in the drawing for bonding, the work shall be coordinated with Civil Contractor. This is applicable in locations as shown in design drawings.

8. At roof G.I. box 160x80x35mm shall be fixed at 500mm from SSL with fixed of PC116 furse earth point with pre-welded joint. The box shall be tied with reinforcement rod and box cover shall be flush with finish wall 25x3 copper tape run on roof perimeter as shown in shop drawing bonded with non ferrous bonding point. All mechanical i.e. AHU's , pole structure, petal structure etc. shall be bonded with 25x3 copper tape.

9. 25x3 copper tape shall be fixed on top of parapet wall at 1m intervals as shown in shop drawing with Non-ferrous bonding point.

10. Inspection shall be offered for QC verification in stages.

11. Inspection request shall be raised al least 24 hours in advance for consultant's inspection, prior to concrete pour/ cover-up work.

12. Final inspection shall be carried out collecting all the data from previous inspection requests and finally verified for the anticipated results.

## QUALITY CONTROL

- In general, it shall be ensure by site engineer that product manufacturer's recommendations are followed and shall be monitored by QA/QC Engineer. However, the following points shall be ensured, in particular;
  - The appropriate and approved material is used
  - Skilled labor is used for application of the material.
  - Inspection request shall be raised for consultant's inspection.
  - QC inspection shall be carried out as per the installation checklist and manufacture's instructions.

## SAFETY

- Work will commence as per safety regulations laid down in the contract specification and project safety plan.
- Proper safety harness to be used and secured, if required.
- All personal protective equipment shall be used as appropriate according to the nature of job.
- Housekeeping shall be of good standard and all cut lengths and debris shall be removed.

## Code No. : MS-EL-07

# Method Statement
# Pulling & Testing of Electrical Wires
# (Conduits and Trunking)

# METHOD STATEMENT FOR PULLING AND TESTING OF WIRES

## SCOPE AND PURPOSE

This method statement covers the on site Pulling and Testing of Electrical wires in PVC/GI conduits and trunking for small power and lighting and the requirement of checks to be carried out.

## REFERENCE DOCUMENTS

- Approved Shop Drawings
- Approves method statement for installation of PVC/GI conduits, trunking and earthing
- Authority regulations

## GENERAL

Electrical wires shall be enclosed in conduit, trunking and with in short lengths of flexible conduit for final connections to the various items of the equipment. Installation wires will be carried out as per the project specifications and drawings and authority regulations.

## MATERIAL

- PVC insulated wires confirming to BS 6004 and IEC-60227 rated at 450/750V
- Heat resisting cables confirming to BS 6007

## RESPONSIBLE PERSONNEL

- Project Electrical Engineer
- Site Electrical Engineer
- Site Supervisor
- Technicians Qualified in the trade

## METHOD OF PRE INSTALLATION

- Ensure that approved material required to carryout work will be available.
- Prior to commencement of work, area and access will be inspected to confirm that the site is ready to commence the work
- All relevant documentation and material applicable to particular section of works will be checked by site engineer before commencement.
- Physical verification of material will be carried out for any damages prior to taking from stores.
- The site engineer / supervisor will give necessary instruction to tradesman and provide necessary construction / shop drawings.

- The site engineer / supervisor will also check that tools and equipments available are in compliance to contract requirements.
- The site supervisor also explains tradesman regarding safety pre-cautions to be observed.
- The site supervisor and QC engineer will ensure that Calibrated Megger is available at site for testing.

## METHOD OF INSTALLATION

1. Where in single core PVC insulated cables are enclosed with in the conduits, core will be taken that no damage will occur to the cables during their installation.
   Where conduits is to be installed in damp conditions or out doors rubber sealing gaskets will be installed behind besa/ adaptable box lids.
2. Number of cables to be pulled into the conduit will be as per **TABLE 1 & 2**. The sum of all factors for the cables, as given in **TABLE 1** shall not be greater than the factor for the conduits as given in **TABLE 2**.

### TABLE 1 FACTORS FOR SINGLE CORE PVC INSULATED CABLES ENCLOSED IN A CONDUIT

| CONDUCTOR SIZE OF CABLES (mm2) | FACTORS | |
|---|---|---|
| | FOR SHORT RUNS | FOR LONG RUNS OR RUNS WITH BENDS |
| 1.5 | 27 | 22 |
| 2.5 | 39 | 30 |
| 4.0 | 58 | 43 |
| 6.0 | 88 | 58 |
| 10 | 146 | 105 |

### TABLE 2 FACTORS FOR THE CONDUITS

| Type of run conduit size (mm) | | Straight run | | | Run with 1 bend | | | Run with 2 bends | | |
|---|---|---|---|---|---|---|---|---|---|---|
| | | 20 | 25 | 32 | 20 | 25 | 32 | 20 | 25 | 32 |
| Length of run | 2m | (460) | (800) | (1400) | 286 | 514 | 900 | 256 | 463 | 818 |
| | 3m | 460 | 800 | 1400 | 270 | 487 | 857 | 233 | 422 | 750 |
| | 4m | 286 | 514 | 900 | 256 | 463 | 818 | 213 | 388 | 692 |
| | 5m | 278 | 500 | 878 | 244 | 442 | 783 | 196 | 358 | 643 |
| | 6m | 270 | 487 | 857 | 233 | 422 | 750 | 182 | 333 | 600 |
| | 7m | 263 | 475 | 837 | 222 | 404 | 720 | 169 | 311 | 563 |
| | 8m | 256 | 463 | 818 | 213 | 388 | 692 | 159 | 292 | 529 |
| | 9m | 250 | 452 | 800 | 204 | 373 | 667 | 149 | 275 | 500 |
| | 10m | 244 | 442 | 783 | 169 | 358 | 643 | 141 | 260 | 474 |
| | 11m | 238 | 433 | 764 | | - | | | - | |
| | 12m | 233 | 424 | 748 | | - | | | - | |
| | 13m | 228 | 416 | 735 | | - | | | - | |

| | 14m | 223 | 408 | 721 | - | - |
|---|---|---|---|---|---|---|
| | 15m | 218 | 401 | 708 | - | - |

**NOTE: 1.** Short run means a straight run not exceeding 3m long. Long run means straight run exceeding 3m long.

**2.** The conduit factors shown in brackets shall only be used in conjunction with the corresponding cable factors for short runs.

**3.** For cables and ro conduits not indicated in Table 1 & 2 the number of cables drawn in to a conduit shall not exceed space factor of 45%.

3. Before pulling wire into cast in conduits, ensure that nylon draw ropes have been installed and the containment system has been found to be clear of any obstruction.

4. The use of lubricants, grease, graphite or talc will not be used to assist the drawing of cables.

5. Cables of different circuit categories will not be mixed with in the same conduit.

6. While preparing cable ends, ensure that conductor stands are not damaged and the strands are twisted together with pliers to ensure neat and firm connection.

7. While removing the conductor insulation, ensure that no excess exposed conductor shall be left.

8. Each circuit will incorporate a separate protective conductor selected in accordance with IEEE regulations (TABLE 54 G latest). Earth continuity will be maintained according to authority's regulations.

9. Final sub circuits will be installed in continuous lengths and no joints will be permitted along the cable run. The final sub circuits will be wired in the loop – in method and all terminations are made will be accessible.

10. Neutral conductors of lighting will be wired direct to the lighting points and will not pass through switch boxes.

11. Where PVC insulated sheathed cable enters or exit the trunking system the hole will be lined up with rubber grommets or bushes.

12. Where cables are enclosed in the same trunking and connected to different distribution boards, they shall be distinguished by separating the cables by insulating taping at an approximate interval of 2.0M and also with an identification labeling indicating circuit type and reference.

13. Color identification sleeve to denote phase, circuit reference and/or terminal reference and/or terminal number to which it is connected coding of wires will be followed.

## QUALITY CONTROL

- In general, it shall be ensure by site engineer that product manufacturer's recommendations are followed and shall be monitored by QA/QC Engineer. However, the following points shall be ensured, in particular;
  - The appropriate and approved material is used
  - Skilled labor is used for application of the material.
  - Inspection request shall be raised for consultant's inspection.
  - QC inspection shall be carried out as per the installation checklist and manufacture's instructions.

## SAFETY

- Work will commence as per safety regulations laid down in the contract specification and project safety plan.
- Proper safety harness to be used and secured, if required.
- All personal protective equipment shall be used as appropriate according to the nature of job.
- Housekeeping shall be of good standard and all cut lengths and debris shall be removed.

## Code No. : MS-EL-08

# Method Statement
# Commissioning Management

# METHOD STATEMENT FOR COMMISSIONING MANAGEMENT

## SCOPE AND PURPOSE

This "Method Statement' 'describes the methodology regarding planning, organizing and executing, methods to be adopted for Commissioning Management and Verification of MEP services installed at Jumeirah Beach Residence-Sector 4-A Specialist Commissioning validation (duly approved by the consultant) to e appointed to monitor, witness and verify all testing and commissioning of the MEP services.

## REFERENCE DOCUMENTS

- Project Specifications Volume III, Division 15, Section: 15010,15100,15400,16005,16010,16050
- Approved shop drawings (latest revision)
- Approved material submittals
- Local authority regulations (DEWA, CIVIL DEFENCE, etc.)

## GENERAL

The MEP Services that will be subjected to Commissioning management and validation procedures include the following:

- Chilled water supply and return systems
- Air conditioning and ventilation systems
- Car park Ventilation Systems
- Domestic water supply system
- Firefighting sprinkler & FM 200 extinguisher systems
- LV Power & Small power
- Emergency lighting & control systems
- Fire detection and alarm system
- Intercom system
- Diesel Generator
- BMS & Controls
- Intercom System

## MATERIAL

### Approved submittals for
- Specialist for testing and commissio'1mg.
- Approved method statements for commissioning.

## EQUIPMENT

Calibrated Inspection, Measuring and Test Equipment as outlined in Method Statements.

## RESPONSIBLE PERSONNEL

- Specialist Commissioning Team
- Project Engineers (HVAC/ELEC./PLUMBING)
- Manufacturer's authorized representatives( where required)
- Construction In-Charge.
- Site Engineer/Site supervisor.

- QA/QC Inspectors.
- Safety officer
- Site Foremen.
- Tradesmen
- Helpers

## COMMISSIONING MANAGEMENT

### OVERVIEW OF COMMISSIONING

### RESPONSIBILITY:-

1. MEP contractor shall submit prequalification documents for Commissioning Management and Verification to the Consultant for approval.

2. Upon approval of a Specialist Commissioning Company, regular site meetings will be convened, in order to ensure that commissioning of each element proceeds in a logical, systematic and cost effective manner from commissioning design appraisal through installation, site testing, commissioning performance verification, whilst critically satisfying the requirements of overall project programme.

3. Site Meetings, should discuss the following clearly defined activities:

   - Design Commission ability Review
   - Commissioning Method Statements
   - Commissioning logic diagrams and Programmes.
   - Format of final record documentation.
   - Construction installation appraisal and progress assessment.
   - Progress on Pre-commissioning verification
   - Progress on verification of all commissioning works
   - Verification of integrated performance testing.
   - Review of record documentation
   - Test documentation collation.

4. It will be ensured that Specialist Commissioning project team will be experienced enough to carryout commissioning verification activities.

5. Prior to any commissioning and testing taking place, it is essential that a detailed and flexible commissioning plan be available. The initial aim of this stage of works will be to ensure that all necessary provisions have been made in the construction design to allow commissioning to be completed to the requirements of the specification and within the confines of the construction programme.

6. Upon appointment of a specialist commissioning agency, a detailed report will be produced indicating the following:

   - The equipment and system facilities required.
   - The on-site testing procedures required.
   - Testing and commissioning risk assessments.
   - The handover procedures and final witnessing requirements.

### COMMISSIONING PLAN

1. Working method statements for each element of the pre-commissioning, commissioning and performance testing will be made available and submitted to the Consultant for approval during planning phase, prior

to the commencement of any testing. A commissioning Management and Verification matrix is attached and will be followed for this purpose.

2. For Progress monitoring purposes a Commissioning programme will be developed and submitted.

3. Throughout the construction period of the project, the specialist commissioning team will undertake detailed appraisals of the installations, to ensure that the systems contain all necessary features that are required for commissioning and performance of the systems, as well as health and safety considerations for those carryout the commissioning works.

## COMMISSIONING PHASE

1. During pre-commissioning and commissioning stages of the project the role of specialist commissioning team is out lined as follows:

2. Ensuring that all necessary pre-commissioning has been carried out to the agreed methods and standards, and is approved by the verification team. This must be completed prior to the start up of all plant and equipment

3. Ensuring that the commissioning and testing works are carried out in accordance with agreed methods and standards as per the commissioning plan.

4. Verifying and collating record documentation for all commissioning and testing operations.

5. Reporting / liasing to ensure that commissioning progress is incorporated into overall completion monitoring process and handover.

6. Commissioning Verification system will require specialist-commissioning contractor to ensure that all tests offered for demonstration have been completed successfully and full documentation is available fully.

## COMMISSIONING, TESTING AND VERIFICATION

Following clause describes the main elements of test validation PE/CME witnessing to be undertaken.

### CHILLED WATER SYSTEM

- Flushing and chemical cleaning
- Pre-commissioning
- Commissioning of Pumps
- Water balancing

### AIR CONDITIONING AND VENTILLATION

- Pre-commissioning.
- Air Balancing (distribution balancing)
- Performance and function testing of AHD's
- Performance and function testing of Fans.
- Performance and function testing of Car Park Ventilation system

### PLUMBING SERVICES

- Pre-commissioning
- Plant and function performance (Verify Statutory acceptance)

## FIRE FIGHTING SPRINKLER AND FM 200 SYSTEM

- Plant function and performance
- Alarms & monitoring interfaces

## LV DISTRIBUTION AND SMALL POWER

- Installation test (Include Meggering)
- Commissioning tests (Verification of switching control and indication)
- Performance testing (verify statutory acceptance)

## FIRE ALARM SYSTEM

- Device address and operation tests
- Performance testing (verify overall cause and effect testing)- Civil Defence approval required

## MISCEEOUS ELECTRICAL SERVICES

- Air craft warning system (by nominated specialist contractor by client)
- Intercom system
- Telecom / Data system
- Diesel Generator
- Emergency lighting (verify function and statutory demonstrations).
- Earthing / bonding & lightning protection (static verification).
- Security system (by nominated specialist sub contractor by client).
- Swimming pools

## BMS AND CONTROLS

- Wiring and panel installation tests
- Commissioning tests (Final function and software graphics).
- Performance testing (overall emergency and plant performance).

## HANDOVER PHASE

### Client and Statutory Demonstrations

Upon satisfactory completion of commissioning of a particular system or sub-system, the Specialist Commissioning Team will issue the relevant test documentation to the MEP Contractor for onward issue to the consultant and statutory authority as necessary.

## COMMISSIONING AND TEST DOCUMENTATION

The Specialist commissioning agency/ Equipment supplier will provide .test pro-forma for all site wide operations including pre-commissioning, balancing and testing of air and water systems, and testing of electrical distribution system. The same will be submitted along with method statements.

Following successful witness, the final approved commissioning documentation, as approved by the specialist-commissioning agency, will be collated and submitted to main contractor for onward submission to consultant.

MEP contractor for incorporation into final project O & M manuals will retain a copy of it.

## MEETINGS REPORTING AND QUALITY MONITORING SYSTEM

### MEETINGS

The following meetings would form part of the specialist commissioning team role throughout the course of works on site: -

- Site progress meetings- if requested by job clients for coordination purposes.
- Commissioning meetings-attended by in-charge Specialist commissioning team.

### REPORTING

Monthly progress report shall be prepared by the Specialist commissioning I agency in addition to individual work tests, performance demonstrations etc.: -

Detailed elemental report on progress.
Updated commissioning activity schedule.

### QUALITY MONITORING SYSTEM

The Specialist commissioning team in association with MEP contractor will provide a detailed breakdown of each element of commissioning for each service. A database will be maintained, showing the status of testing and commissioning process, giving all levels of site management a clear and concise breakdown status.

Specialist commissioning team, for comments on site installation and testing will provide site observation reports.

Testing and commissioning documentation will be compiled monitored in generated database format.

<u>Code No. : MS-EL-09</u>

# Method Statement
# BUS BAR TRUNKING SYSTEM

www.arabmep.com

# METHOD STATEMENT FOR BUS BAR TRUNKING SYSTEM

## SCOPE AND PURPOSE

This "Method Statement" covers the on site installation and testing of Bus bar Trunking systems with relevant accessories and the requirements of checks to be carried out.

## REFERENCE DOCUMENTS

- Project Specifications Volume III, Division 16, Section: 16010, 16050, 16477
- Approved shop drawings (latest revision)
- Approved material submittals
- Manufacturer's installation instruction
- Local authority regulations (DEWA, CIVIL DEFENCE, etc.)

## GENERAL

Bus - bar trunking system generally includes standard straight lengths, special straight lengths (wherever necessary), elbows (vertical and horizontal), offsets, T -joints, standard component tap off boxes, feeder boxes, fixing elements, flanges, support systems and end connections etc; various items of the equipment. Installation of bus ducts will be carried out as per the project specifications and drawings and DEWA regulations.

## EQUIPMENT

- Electrician hand tools
- Torque wrenches
- Scaffoldings
- Nylon slings
- Measuring Tape
- Calibrated Megger

## RESPONSIBLE PERSONNEL

- Project Engineers
- Construction In-Charge.
- Site Engineer/Site supervisor.
- QA/QC Inspectors.
- Safety officer
- Site Foremen.
- Tradesmen
- Helpers

## METHOD OF PRE INSTALLATION

- Ensure that approved material required to carryout work will be available.
- Bus bars complete with accessories received at site will be inspected as per approved material submittal and manufacturer's approved isometric drawings for completeness and physical damages if any. In case of any damage, the same should be brought to the notice of supplier for suitable resolution/replacement.

| | |
|---|---|
| PLANNING DEPARTMENT | |
| Rev.: 0 | **MEP PLANNING MANUAL** |
| Date of Issue: Feb 2007 | |
| Doc No.: MEP-01-07 | |

- Physical verification of materials will be carried out for any damages prior to taking from stores and also prior to installation. Manufacturer's test certificates received will be reviewed and submitted to consultant for approval.
- Prior to commencement of work, area and access will be inspected to confirm that the site is ready to commence the work
- All relevant documentation and material applicable to particular section of works will be checked by site engineer before commencement.
- Physical verification of material will be carried out for any damages prior to taking from stores.
- The site engineer / supervisor will give necessary instruction to tradesman and provide necessary construction / shop drawings.
- The site engineer / supervisor will also check that tools and equipments available are in compliance to contract requirements.
- The site supervisor also explains tradesman regarding safety pre-cautions to be observed.

## METHOD OF INSTALLATION

1.  Bus bar installation will be carried out as per approved bus bar lay out/isometric drawings. Before starting the installation of bus bar system, proper co ordination shall be done with other trades.
2.  Manufacturer's installation instructions attached will be followed for handling, storage, installations and testing.
3.  Determine the position of the bus bar supports as per approved construction drawings and mark them on concrete surfaces.
4.  Fix the support as per manufacturer's installation instructions.
5.  Provide sufficient horizontal and vertical clearance from walls and ceilings to provide easy access to joints, both for permanent installation and possible future removal of section when required.
6.  For vertical bus bars, ensure that no joints or expansion joints are installed in the slab thickness.
7.  Bus bar will be leveled and mounted at the correct height.
8.  Elbows, offset and tap off boxes etc., will be installed as per approved construction drawings and as per manufacturer's instructions.
9.  Cable gland plates for tap off boxes and fixing elements shall be installed as per the recommendations of the Manufacturer and as per approved schematics.
10. Cable gland plates for tap off boxes and fixing elements shall be installed as per the recommendations of the Manufacturer and as per approved schematics.
11. Bus bar should be installed so that the orientations of phase from the front side will be GRY&N. The cabling from tap off shall be from side or bottom.

### *JOINT ASSEMBLY*

12. Ensure that all contact surfaces are clean and free of containments.
13. Align the bus bar ends of adjoining sections, verifying proper phase alignment, and slide the sections together (use joints pullers, if required).

14. Expansion joints if required as shown in approved construction drawings will be installed as manufacturer's instructions attached.

15. After completing the jointing of assemblies, torque the joints bolts to the specific value.

16. Where a bus bar extends through a inside block. Shear walls, an internal transverse barrier supplied by manufacturer will be installed.

17. Bus ways to be properly covered during the installation to protect them from moisture or other types of contaminants.

18. All the floor/wall crossing of bus bars is to sealed with approved type of fire sealant.

19. Bus bar enclosure continuity shall be ensured with proper bounding and connected to the main earthing grid through return paths al LV Panel end.

## *TESTING*

After complete installation of Bus bar systems, Bus bars are subjected to following tests as recommended by Manufacturer:

- Checking of all joint connections for tightness.
- An Insulation Resistance test, carried out with a calibrated Megger insulation tester, to ensure that system is free from short circuits and grounds (phase-to-ground, phase -to- neutral and phase-to-phase). It should be noted that readings vary inversely with the length of run and width or number of bars per phase.
- Verify that system phasing matches the busway phasing before reconnecting all connections to switchboards.
- Continuity tester will be used to check electrical continuity of SS/QCE phase bus bars, neutral bus bar and earth bus bar.

## QUALITY CONTROL

- QCE along with Project Engineer and site Supervisor will monitor that all components are installed as per the contract specifications         and approved submittals.
- Inspection Request (IR) shall be submitted to the Main PE/SS/QCE Contractor/Consultant during the following stages :-
- After completion of installation of bus ducts for a particular area/section.
- For testing of Bus ducts.

## SAFETY

- Work will commence as per safety regulations laid down in the contract specification and project safety plan.
- Proper safety harness to be used and secured, if required.
- All personal protective equipment shall be used as appropriate according to the nature of job.
- Housekeeping shall be of good standard and all cut lengths and debris shall be removed.
- Hoisting and handling of bus ducts will be carried out as per manufacturer's installations.

## Code No. : MS-EL-10

# Method Statement
# Building Management System

# METHOD STATEMENT FOR BUILDING MANAGEMENT SYSTEM

## SCOPE AND PURPOSE

This method statement covers the site wiring, termination and commissioning of Building management system installed.

## REFERENCE DOCUMENTS

- Approved Shop Drawings
- Approves Material Submittals

## GENERAL

The work station controls/ monitor equipments such as Pumps, AHD's / FCU's etc; through cable network, FCU/DDC controllers and various peripheral instruments installed in the system as per approved layouts.

## EQUIPMENT / TOOLS

- Calibrated Megger
- Calibrated Multi meter
- Operator terminal (PXM20)
- Electrician hand tools
- Lap top computer

## RESPONSIBLE PERSONNEL

- Commissioning Engineer
- Project Electrical Engineer
- Site Electrical Engineer
- QA / QC Engineer
- Site Supervisor
- Electrician
- Helpers

## PROCEDURE

### Wiring, termination and pre-commissioning procedures:

1. Ensure that locations all Peripheral devices installed in various systems are as per approved layout drawings and suitable for system operation logics.

| | |
|---|---|
| PLANNING DEPARTMENT | |
| Rev.: 0 | **MEP PLANNING MANUAL** |
| Date of Issue: Feb 2007 | |
| Doc No.: MEP-01-07 | |

2. Ensure that installations of DDC panel/FCU Controllers are complete and are acceptable.

3. Ensure that all cable containment and cabling is completed as per approved drawings and As-Built marked up drawings are available reflecting the actual site installations.

4. Siemens will be notified, when the installations are complete to ensure that installations are acceptable.

5. Terminations to DDC panels/ FCU controllers will be carried out as per latest approved drawings and documents.

6. Continuity checks for wiring (for DDC/FCU loops) will be carried and the same are acceptable.

7. Prior to Commencement of Commissioning, areas and access will be inspected to confirm that Site is ready to commence the commissioning.

8. Ensure that all instruments are calibrated and are in proper working condition.

**Method of Commissioning**

1. Commissioning of Building management system will be carried out as per SIEMENS Method statements attached.

   - MS001- Wiring and termination of DDC Panels
   - MS002- Wiring and termination of LON FCU Controllers
   - MS003- Commissioning of Outstation panels(DDC Panel)
   - MS004- Commissioning of FCU Controller
   - MS005- Commissioning of Central interface panel & BMS Management station.

2. All the test documentation will be recorded on the test sheets attached to the above Siemens Procedures attached.

## QUALITY

- QCE in coordination with Commissioning Manager and SIEMENS Engineer notify to Consultants for Wire termination and continuity checks for each building/area.

- QCE in coordination with Commissioning Manager and SIEMENS Engineer notify to Consultants for commissioning of DDC outstation panel, FCU Controller and Central inter face panel & BMS management station.

- QCE in association with Commissioning Manager and Project Engineer will ensure that all the test documentation is complete and signed off.

| PLANNING DEPARTMENT | |
| --- | --- |
| Rev.: 0 | |
| Date of Issue: Feb 2007 | **MEP PLANNING MANUAL** |
| Doc No.: MEP-01-07 | |

## SAFETY

- Work will commence as per safety regulations laid down in the contract specification and project safety plan.
- Proper safety harness to be used and secured, if required.
- All personal protective equipment shall be used as appropriate according to the nature of job.
- Housekeeping shall be of good standard and all cut lengths and debris shall be removed.

## ATTACHMENTS

**SIEMENS METHOD STATEMENT FOR BUILDING MANAGEMENT SYSTEM ALONG WITH TYPICAL CHECKLIST ATTACHED.**

# Section 10

# Method Statement for Mechanical (HVAC) Installation

## Code No. : MS-AC-01

# Method Statement
# Installation of HVAC Ductwork

# METHOD STATEMENT FOR INSTALLATION OF HVAC DUCTWORK

## SCOPE AND PURPOSE

This "Method Statement" covers the on site installation of HVAC duct work and accessories and the requirement of checks to be carried out.

## REFERENCE DOCUMENTS

- Project Specifications
- Approved shop drawings (latest revision)
- Approved material submittals
- Duct work standard DW 144

## GENERAL

HVAC duct f work generally includes all types of Supply Air Duct, Return Air Duct, Fresh and Exhaust Air Duct and their accessories such as duct elbows, offsets, transformation pieces branch off pieces, tee connections, access doors, Fire Dampers, Volume Control Dampers, Sound Attenuators, flexible ducting and insulation.

## EQUIPMENT

- Portable grinding Machine
- Drilling Machine
- Sheet metal cutting tools and bender
- Spirit level
- Scaffolding
- Duct erector hand tools
- Testing instruments
- Air Compressor

## RESPONSIBLE PERSONNEL

- Project Engineers
- Construction In-Charge.
- Site Engineer/Site supervisor.
- QA/QC Inspectors.
- Safety officer
- Site Foremen.
- Pipe fitters / Welders
- Helpers

## METHOD OF PRE INSTALLATION

- Ensure that approved material required to carryout work will be available.

- Prior to commencement of work, area and access will be inspected to confirm that the site is ready to commence the work
- All relevant documentation and material applicable to particular section of works will be checked by site engineer before commencement.
- Physical verification of material will be carried out for any damages prior to taking from stores.
- The site engineer / supervisor will give necessary instruction to tradesman and provide necessary construction / shop drawings.
- The site engineer / supervisor will also check that tools and equipments available are in compliance to contract requirements.
- The site supervisor also explains tradesman regarding safety pre-cautions to be observed.
- Prior to Leak testing, Site Engineer will ensure that Calibrated test kit is available and are in good condition.

## METHOD OF INSTALLATION

1. Prior to commencement of work coordination will be done with other
2. Determine the position of duct supports as per approved construction layouts and specification.
3. Prepare and fix the duct supports as per approved construction drawing and specification.
4. Any cut edges of angles, channels or threaded rods will be touch up with Zinc rich paint.
5. Transport the Duct pieces and fittings to final location.
6. Pre-assemble the Duct pieces ''and ' fittings as per approved shop drawing ensuring the alignment.
7. Acoustic insulation will be carried out wherever required.
8. Raise the duct work on to the supports ensuring that each length is aligned with preceding length as per dimensions shown on approved shop drawings.
9. Approved duct sealant shall be applied on the joints. Any excess sealant so that the joint left in clean and tidy condition.
10. Ensure that duct work is clean and no tools/ construction debris exists within duct work before proceeding to next length.
11. All open ends of the duct works shall be temporarily sealed with polythene sheets/ply wood before leaving the job site to prevent moisture and dirt.
12. Ensure that all accessories like Volume control dampers, Fire dampers, Access doors, Test points, Sensors are installed in accordance with approved shop drawings.
13. Install Sound Attenuators according to approved shop drawings
14. Installation of duct work (complete with accessories) shall be checked before applying insulation at joints.
15. Leak test will be carried out for duct work as applicable in DW144 standard.
16. Insulation of duct work will be completed as per manufacturer's recommendations (copy-enclosed).
17. Ensure the duct surface is clean and dry before applying any insulation material.
18. Ensure the thickness of insulation as per approved shop drawing.

| PLANNING DEPARTMENT | |
| --- | --- |
| Rev.: 0 | |
| Date of Issue: Feb 2007 | **MEP PLANNING MANUAL** |
| Doc No.: MEP-01-07 | |

19. Ensure that all edge joints are closely butted and ends are flush and seated properly.

20. Apply ALUGLASS Tape (Self Adhesive Aluminium Foil Laminated on Glass fabric) at Joints of insulation.

21. Ensure the continuity of vapor barrier and other protective coatings on insulation surface as well as at connections.

22. Ensure firm adherence of insulation around_ ducting by using approved adhesive between sheet metal and Insulation material.

23. Where insulated duct work passes through fire rated wall/partitions the gap between sleeve and duct work shall be filled with approved fire barrier.

## QUALITY CONTROL

- QCE along with Project Engineer and site Supervisor will monitor that all components are installed as per the contract specifications and approved submittals.
- Inspection Request (IR) shall be submitted to the Main Contractor/Consultant during the following stages :-
  - After completion of installation before testing
  - Leak testing of duct work
  - After complete insulation

## SAFETY

- Work will commence as per safety regulations laid down in the contract specification and project safety plan.
- Proper safety harness to be used and secured, if required.
- All personal protective equipment shall be used as appropriate according to the nature of job.
- Housekeeping shall be of good standard and all cut lengths and debris shall be removed.
- Good ventilation for duct work insulation shall be ensured.

## Code No. : MS-AC-02

# Method Statement
# Installation and insulation of Spiral Round Ducting and Duct Accessories

| PLANNING DEPARTMENT | |
| --- | --- |
| Rev.: 0 | |
| Date of Issue: Feb 2007 | **MEP PLANNING MANUAL** |
| Doc No.: MEP-01-07 | |

# METHOD STATEMENT FOR INSTALLATION AND INSULATION OF SPIRAL ROUND DUCTING AND DUCT ACCESSORIES

## SCOPE AND PURPOSE

This "Method Statement" covers the on site installation of Spiral round ducting and duct accessories and the requirement of checks to be carried out.

## REFERENCE DOCUMENTS

- Project Specifications
- Approved shop drawings (latest revision)
- Approved material submittals
- Duct work standard DW 144

## GENERAL

GI Spiral round ducting generally installed in for FCU return air, and supply air as shown in approved shop drawing includes spiral round ducting, fittings such as elbows, tees, couplings, reducers, collars etc, volume control dampers, flexible ducting insulation.

## EQUIPMENT

- Portable grinding Machine
- Drilling Machine
- Sheet metal cutting tools and bender
- Spirit level
- Scaffolding
- Duct erector hand tools
- Testing instruments
- Air Compressor

## RESPONSIBLE PERSONNEL

- Project Engineers
- Construction In-Charge.
- Site Engineer/Site supervisor.
- QA/QC Inspectors.
- Safety officer
- Site Foremen.
- Pipe fitters / Welders
- Helpers

## METHOD OF PRE INSTALLATION

- Ensure that approved material required to carryout work will be available.

- Prior to commencement of work, area and access will be inspected to confirm that the site is ready to commence the work
- Check all the GI Spiral ducting received are pre fabricated with tag number labeled and received at site in accordance with specifications and correct dimensions as per approved drawings.
- All relevant documentation and material applicable to particular section of works will be checked by site engineer before commencement.
- Physical verification of material will be carried out for any damages prior to taking from stores.
- The site engineer / supervisor will give necessary instruction to tradesman and provide necessary construction / shop drawings.
- The site engineer / supervisor will also check that tools and equipments available are in compliance to contract requirements.
- The site supervisor also explains tradesman regarding safety pre-cautions to be observed.

## METHOD OF INSTALLATION

1. Prior to commencement of work coordination will be done with other services.
2. Determine the position of duct supports as per approved Shop drawings and coordinated layouts and specification.
3. Prepare and fix the duct supports as per approved construction drawing and specification.
4. Any cut edges of angles, channels or threaded rods will be touch up with Zinc rich paint.
5. Transport the duct pieces and fittings to final location.
6. Before assembly ensure that all ducts are free from dirt.
7. Check that ducts and fittings are undamaged. This is specially important with regard to the rubber gaskets.
8. Assemble the Duct pieces and fittings as per approved shop drawing ensuring the alignment.
   a) Push the fittings into the duct right to the stop. Turning the fitting a little makes insertion easier.
   b) Fasten fittings to the duct with self tapping screws or centered pop rivets.

   c) Distribute the screws or pop rivets evenly around the circumference, ensuring the rubber gaskets are not damaged i.e. placing the them appox.l0mm from stop and end of the duct. In the event of incorrect assembly, holes caused by screws or pop rivets must be sealed.
9. Raise the duct work on to the supports ensuring that each length is aligned and leveled with preceding length as per dimensions shown on approved shop drawings.
10. Ensure that duct work is clean and no tools/ construction debris exists within duct work before proceeding to next length.
11. All open ends of the duct works shall be temporarily sealed with polythene
    Sheets/ply wood before leaving the job site to prevent moisture and dirt.
12. Ensure that all accessories like Volume control dampers, Test points, Sensors are installed in accordance with approved shop drawings.
13. Installation of duct work (complete with accessories) shall be checked before applying insulation at joints.
14. Insulation of duct work will be completed as per manufacturer's recommendations (copy enclosed).

| PLANNING DEPARTMENT | |
| --- | --- |
| Rev.: 0 | |
| Date of Issue: Feb 2007 | **MEP PLANNING MANUAL** |
| Doc No.: MEP-01-07 | |

a) Ensure the duct surface is clean and dry before applying any insulation material. Apply thinner /cleaner where necessary to make the area grease free.

b) Apply glue recommended by manufacturer with an even spread on complete surface of insulation sheet.

c) Once the glue on the sheet gets dry, apply glue on the GI duct and let it dry, then stick the sheet on one end and slowly press the sheet on the duct from one end to the other so as to ensure that the sheet sticks on the GI duct completely avoiding air bubbles between the insulation sheet and GI duct.

d) In the areas where the GI duct comes into contact with the duct hangers, rigid support is recommended such as wood. The wood piece should be glued onto the duct hanger and layer of rubber to be glued on the wood, so that there is no direct contact between the duct insulation and the wooden piece thus preventing tearing of the duct insulation. In case the positioning of the support is available in advance, wood by itself can be used. In this case it has to be made sure that insulation will be glued to the wood from both sides accordingly.

15. Where insulated duct work passes through fire rated wall/partitions the gap between sleeve and duct work shall be filled with approved fire barrier.

## QUALITY CONTROL

QCE along with Project Engineer and site Supervisor will monitor that all components are installed as per the contract specifications and approved submittals.

- Inspection Request (IR) shall be submitted to the Main Contractor/Consultant during the following stages :-
- Mock up installation of GI spiral round ducting before and after insulation
- After complete insulation

## SAFETY

- Work will commence as per safety regulations laid down in the contract specification and project safety plan.
- Proper safety harness to be used and secured, if required.
- All personal protective equipment shall be used as appropriate according to the nature of job.
- Housekeeping shall be of good standard and all cut lengths and debris shall be removed.

## Code No. : MS-AC-03

# Method Statement
# Installation, Testing & Insulation of Chilled Water System

# METHOD STATEMENT FOR INSTALLATION, TESTING & INSULATION OF CHILLED WATER SYSTEM

## SCOPE AND PURPOSE

This "Method Statement" covers the on site installation, testing and insulation of the chilled water piping system including risers and the requirements of checks to be carried out.

## REFERENCE DOCUMENTS

- Project Specifications
- Approved shop drawings (latest revision)
- Approved material submittals

## GENERAL

Chilled water piping system includes chilled water pipes, fittings, valves and accessories used for transportation (supply and return) chilled water for AHU's and FCU's through Chilled water pumps. Pipes and fittings up to 50 NB dia. size shall be threaded type and 65 NB and above shall be either grooved or welded type.

## EQUIPMENT

- Electrical grinding Machine
- Scaffolding
- Drilling Machine
- Ladders
- Heavy duty cutter.
- Pipe fitter hand tools
- Grooving machine
- Electrode oven/ Quivers
- Threading machine
- Pressure test pump
- Welding machine
- Test pressure gauges
- Cutting Torch set.

## RESPONSIBLE PERSONNEL

- Project Engineers
- Construction In-Charge.
- Site Engineer/Site supervisor.
- QA/QC Inspectors.
- Safety officer
- Site Foremen.
- Pipe fitters / Welders
- Helpers

## METHOD OF PRE INSTALLATION

- Ensure that approved material required to carryout work will be available.

- Proper materials including chilled water pipes, fittings and associated accessories are with drawn from stores according to approved shop drawing and good engineering practices.

- Prior to commencement of work, area and access will be inspected to confirm that the site is ready to commence the work

- All relevant documentation and material applicable to particular section of works will be checked by site engineer before commencement.

- Physical verification of material will be carried out for any damages prior to taking from stores.

- The site engineer / supervisor will give necessary instruction to tradesman and provide necessary construction / shop drawings.

- The site engineer / supervisor will also check that tools and equipments available are in compliance to contract requirements.

- The site supervisor also explains tradesman regarding safety pre-cautions to be observed.

- Prior to Hydrostatic Pressure testing, Site Engineer will ensure that Calibrated pressure gauges are available and are in good condition

## METHOD OF INSTALLATION

1. All welding activities will be carried out by certified welders only.

2. Supervisor/Foremen will carryout a site survey and mark the route of Chilled water piping as per approved shop drawings. In the event that there are any discrepancies or difficulties in executing the work, these will be brought to the notice of Project Engineer for corrective action.

3. Determine the position of supports and fix the supports using anchor bolts and ensure all fixing are tight and secure.

4. Any cut edges of angles, channels or threaded rods will be touch up with Zinc rich paint.

5. Install the pipes in position by using suitable lifting equipments( If necessary).

6. Assemble the pipes and fittings as per approved shop drawing.

7. After installation of pipe work check for correct leveling, position alignment and proper grooving/threading or welding.

8. Where the pipes of dissimilar materials are to be joined together necessary; dielectric unions shall be used.

9. Sufficient spacing shall be maintained between pipes for insulation.

10. Spacing between supports / hangers will be maintained in accordance with latest approved shop drawings.

11. Ensure all open ends of pipes, fittings and valves are covered with polyethylene sheet before leaving work space.

12. All high point on piping system will be provided with an air vent. Drains ; will be provided at low point with an access. High point vents will be connected nearest drains.

13. **INSTALLATION OF VALVES AND ACCESSORIES**

    - Install system valves and accessories as per latest approved shop drawings.

    - Ensure that system equipment, valves and accessories are secure and rigid

- The installation shall be done allowing sufficient access to all Valves/strainers/Gauges as per Manufacturer's recommendations.

## 14.    INSTALLATION OF CHILLED WATER RISERS

- Pipe sizes will be identified first as per latest approved shop drawing and shifted to respective floors.
- Install the supports as per approved shop drawing.
- The pipe shall be thoroughly cleaned prior to joining.
- On completion of joining, install the pipes using necessary equipment / manpower.
- After installation of risers check the pipeline for proper alignment and supports.

## HYDROSTSTIC PRESSURE TESTING

1. Complete pipe work will be subjected to hydraulic pressure tested as per technical specification. Depending on ongoing Construction activities sectional hydro testing will be under taken to meet the requirements of the programme. Test pressure will not be less than 1.5 times the working pressures but not less than 1035 KPa (for 24 hour period) which ever is greater. Prior to any testing the system pressure will be detailed on the pressure testing documentation.
2. Pressure gauges used for Pressure testing will have valid calibration certificate.
3. After successful Pressure testing ensure that piping system is fully drained and released for chemical cleaning which will be carried out at later date as per approved method statement

## INSULATION

1. Before application of thermal insulation, Chilled water pipes will be ; painted with a primer paint as per specification. Painting of welded joints will be carried out after pressure testing.
2. Insulation of chilled water pipe work will be carried out as per details as shown in approved submittals. Thermal insulation of welded joints will be carried out after pressure testing.
3.    Ensure thickness of insulation is as per approved drawing

## QUALITY CONTROL

- QCE along with Project Engineer and site Supervisor will monitor that all components are installed as per the contract specifications        and approved submittals.
- Inspection Request (IR) shall be submitted to the Main Contractor/Consultant during the following stages :-
- After completion of installation before hydrostatic pressure test.
- Pressure testing of piping
- After completion of insulation

## SAFETY

- Work will commence as per safety regulations laid down in the contract specification and project safety plan.
- Proper safety harness to be used and secured, if required.
- All personal protective equipment shall be used as appropriate according to the nature of job.
- Housekeeping shall be of good standard and all cut lengths and debris shall be removed.
- Fire Extinguishers will be provided in the near vicinity during welding and cutting operations.
- Where ever required fire blanket will be provided.
- Hot work permit system will be followed.

## Code No. : MS-AC-04

# Method Statement
# Installation of Fan Coil Unit (FCU)

# METHOD STATEMENT FOR INSTALLATION OF FAN COIL UNIT

## SCOPE AND PURPOSE

This "Method Statement" covers the on site installation of FAN COIL UNITS (FCU) and the requirements of checks to be carried out.

## REFERENCE DOCUMENTS

- Project Specifications
- Approved shop drawings (latest revision)
- Approved material submittals

## GENERAL

Fan Coil Units generally be installed in locations shown in drawings, serving apartments mechanical -rooms etc.-; both yin concealed areas and exposed supplying cold dehumidified air conditioned space.

## EQUIPMENT

- Drilling Machine
- Spirit Level
- Scaffolding
- Hand tools of Trads men

## RESPONSIBLE PERSONNEL

- Project Engineers
- Construction In-Charge.
- Site Engineer/Site supervisor.
- QA/QC Inspectors.
- Safety officer
- Site Foremen.
- Pipe fitters
- Helpers

## METHOD OF PRE INSTALLATION

- Ensure that approved material required to carryout work will be available.
- Prior to commencement of work, area and access will be inspected to confirm that the site is ready to commence the work
- All relevant documentation and material applicable to particular section of works will be checked by site engineer before commencement.
- Physical verification of material will be carried out for any damages prior to taking from stores.
- The site engineer / supervisor will give necessary instruction to tradesman and provide necessary construction / shop drawings.

- The site engineer / supervisor will also check that tools and equipments available are in compliance to contract requirements.
- The site supervisor also explains tradesman regarding safety pre-cautions to be observed.

## METHOD OF INSTALLATION

1. Prior to commencement of work coordination will be done with other services.
2. Determine the position of Fan Coil Unit on the ceiling and mark the location of supports as per approved shop drawing.
3. Prepare and fix the Fan Coil Unit supports as per approved construction drawing and specification.
4. Ensure that Vibration Isolators of approved make, type and model are installed.
5. Install the Fan Coil Units by lifting it slowly by using suitable lifting aids (if necessary). Manufacturer's recommendations shall be followed during installation.
6. Any cut edges of angles, channels or threaded rods will be touch up with Zinc rich paint.
7. Fan Coil Units will be connected with Piping Connections complete with valves and accessories as indicated in approved shop drawing.
8. Ensure that Dielectric unions are used for piping connection to FCU's.
9. FCU Valve packages shall be provided with drain pan as shown in approved shop drawing
10. Ensure that drain connections are made with adequate slope with running trap.
11. Complete the duct connections to Fan Coil Units as shown in approved shop drawing.
12. Complete the Electrical power connection includes earthing a all respects as per approved electrical drawing and Manufacturer's recommendations.
13. Ensure that adequate space for maintenance of fan coil units and valve package is available.
14. Install the thermostat control units as per shop drawing/ Architectural PE/SS/FM drawing.
15. Complete the BMS interfacing with DDC controllers as per approved BMS drawings.

## QUALITY CONTROL

- QCE along with Project Engineer and site Supervisor will monitor that all components are installed as per the contract specifications   and approved submittals.
- Inspection Request (IR) shall be submitted to the Main Contractor/Consultant.

## SAFETY

- Work will commence as per safety regulations laid down in the contract specification and project safety plan.
- Proper safety harness to be used and secured, if required.
- All personal protective equipment shall be used as appropriate according to the nature of job.
- Housekeeping shall be of good standard and all cut lengths and debris shall be removed.

## Code No. : MS-AC-05

# Method Statement
# Installation of Fans

| PLANNING DEPARTMENT | |
| --- | --- |
| Rev.: 0 | |
| Date of Issue: Feb 2007 | **MEP PLANNING MANUAL** |
| Doc No.: MEP-01-07 | |

# METHOD STATEMENT FOR INSTALLATION OF FANS

## SCOPE AND PURPOSE

This "Method Statement" covers the on site installation of FANS and the requirements of checks to be carried out.

## REFERENCE DOCUMENTS

- Project Specifications
- Approved shop drawings (latest revision)
- Approved material submittals

## GENERAL

Fans generally are installed in locations shown in drawings. Types of fans used in mainly classified as Toilet extract fans(TEF),Kitchen extract fans(KEF),Refuse fan s, Lobby pressure relief fans(LRF),Stairwell pressurization fans(SPF),Car park exhaust fans, Jet fans etc.

## EQUIPMENT

- Drilling Machine
- Spirit Level
- Lifting Equipments (Cranes)
- Scaffolding
- Hand tools of Trades men

## RESPONSIBLE PERSONNEL

- Project Engineers
- Construction In-Charge.
- Site Engineer/Site supervisor.
- QA/QC Inspectors.
- Safety officer
- Site Foremen.
- Pipe fitters
- Helpers

## METHOD OF PRE INSTALLATION

- Ensure that approved material required to carryout work will be available.
- Check the name plate details of Fans as per approved shop, drawing/schedules and material submittals before installation.
- Prior to commencement of work, area and access will be inspected to confirm that the site is ready to commence the work
- All relevant documentation and material applicable to particular section of works will be checked by site engineer before commencement.

- Physical verification of material will be carried out for any damages prior to taking from stores.
- The site engineer / supervisor will give necessary instruction to tradesman and provide necessary construction / shop drawings.
- The site engineer / supervisor will also check that tools and equipments available are in compliance to contract requirements.
- The site supervisor also explains tradesman regarding safety pre-cautions to be observed.

## METHOD OF INSTALLATION

1. Prior to commencement of work coordination will be done with other services.
2. Make sure that fans are free from damage and all internal components are complete and in good condition.
3. Fan assemblies will be transported to the nearest point of erection. Care will be taken while handling the units to avoid damage/distortion.
4. Manufacturer's recommendation will be followed for installation of fans.
5. Fans will be installed in location as per approved shop drawing.

## INSTALLATION OF FLOOR MOUNTED FANS

1. Ensure that builders work foundation is provided as per approved shop drawing.
2. Ensure the level of Foundation by spirit level.
3. Check the size and orientation of foundation for its suitability to install the fans.
4. Fix the Vibration Isolators to foundation as per approved submittal.
5. Install the Fan assembly mounting brackets on vibration Isolators as per manufacturer's recommendations.
6. Complete the ductwork / damper installation as per approved shop drawing.

## INSTALLATION' OF IN "LINE MOUNTED FANS(CEILING SUSPENDED)

1. Fans can be mounted either horizontally or vertically as per approved shop drawing.
2. Support the fans by using threaded rods to the fan casing as per manufacturer's recommendations.
3. Provide vibration isolators (approved type) as per manufacturer's recommendations on the mounting brackets/holes.
4. Complete the ductwork/damper connection as per approved shop drawing.
5. Ensure that sufficient space is available to allow removal of access covers and subsequent removal of fan and motor assemblies etc. as per manufacturer's recommendations.
6. Complete all electrical connections as per approved electrical drawing and manufacturer's terminal diagram.
7. Earthing shall be provided as per requirements.
8. Complete the labeling of electrical connections as per schematic drawings.
9. Fan rotation shall be checked before duct connection.

| PLANNING DEPARTMENT | |
| --- | --- |
| Rev.: 0 | **MEP PLANNING MANUAL** |
| Date of Issue: Feb 2007 | |
| Doc No.: MEP-01-07 | |

## QUALITY CONTROL

- QCE along with Project Engineer and site Supervisor will monitor that all components are installed as per the contract specifications and approved submittals.
- Inspection Request (IR) shall be submitted to the Main Contractor/Consultant.

## SAFETY

- Work will commence as per safety regulations laid down in the contract specification and project safety plan.
- Proper safety harness to be used and secured, if required.
- All personal protective equipment shall be used as appropriate according to the nature of job.
- Housekeeping shall be of good standard and all cut lengths and debris shall be removed.
- All lifting operations shall be monitored by Safety Officer.

## Code No. : MS-AC-06

# Method Statement
# Installation of Chilled Water Pumps

# METHOD STATEMENT FOR INSTALLATION OF CHILLED WATER PUMPS

## SCOPE AND PURPOSE

This "Method Statement" covers the on site installation of CHILLED WATER PUMPS and the requirements of checks to be carried out.

## REFERENCE DOCUMENTS

- Project Specifications
- Approved shop drawings (latest revision)
- Approved material submittals

## GENERAL

Chilled water Pumps generally to be installed in locations shown in drawings in the Plant Rooms.

## EQUIPMENT

- Drilling Machine
- Spirit Level
- Lifting Equipments (Cranes)
- Scaffolding
- Hand tools of Trades men
- Alignment tools

## RESPONSIBLE PERSONNEL

- Project Engineers
- Construction In-Charge.
- Site Engineer/Site supervisor.
- QA/QC Inspectors.
- Safety officer
- Site Foremen.
- Pipe fitters
- Helpers

## METHOD OF PRE INSTALLATION

- Ensure that approved material required to carryout work will be available.
- Check the name plate details of Chilled Water Pump as per approved shop, drawing/schedules and material submittals before installation.
- Prior to commencement of work, area and access will be inspected to confirm that the site is ready to commence the work

- All relevant documentation and material applicable to particular section of works will be checked by site engineer before commencement.
- Physical verification of material will be carried out for any damages prior to taking from stores.
- The site engineer / supervisor will give necessary instruction to tradesman and provide necessary construction / shop drawings.
- The site engineer / supervisor will also check that tools and equipments available are in compliance to contract requirements.
- The site supervisor also explains tradesman regarding safety pre-cautions to be observed.

## METHOD OF INSTALLATION

1. Prior to commencement of work coordination will be done with other services.
2. Make sure that the Pumps are free from damage and all internal components are complete and in good condition.
3. Chilled Water Pumps assemblies will be transported to the nearest point of erection. Care will be taken while handling the units to avoid Damage/distortion.
4. Ensure that foundations of Chilled Water Pumps are as per approved shop drawing.
5. Install Inertia base assembly on the foundation as per approved shop drawing. Ensure the oriental' n, axis and level of inertia base is as per approved shop drawing.
6. Fix the vibration isolators to inertia base as per approved shop drawing. Ensure that the locations are matching with Chilled Water Pump base anchoring details.
7. Complete the concrete filling in the Inertia base as per approved shop drawing.
8. Concreting of Inertia bases will be carried out by Main Contractor.
9. Install the chilled water pumps on the inertia base with vibration isolators connected to chilled water pump base frame.
10. Ensure the orientation axis and level of pump as per approved shop drawing.
11. Ensure that pump and motors are properly aligned.
12. Complete the piping connections, including Valves and accessories/flexible connections to pump suction and discharge sides as per approved shop drawing.
13. Piping connections shall be erected to allow access for operation and maintenance of pump motor and valves.
14. Make sure that the piping connections are supported properly and no imposed load of piping is transferred to the pump.
15. Complete all instrument mountings required as approved shop drawing.
16. Complete all electrical connections to pump motor with all necessary electrical protection and controls as per approved electrical shop drawing.
17. Complete the final alignment of pump and motor under manufacturer's representative's supervision.

## QUALITY CONTROL

- In general, it shall be ensured by Site Engineer that product manufacturer's recommendations are

| PLANNING DEPARTMENT | |
| --- | --- |
| Rev.: 0 | |
| Date of Issue: Feb 2007 | **MEP PLANNING MANUAL** |
| Doc No.: MEP-01-07 | |

followed and shall be monitored by QA/QC Engineer. However, the following points shall be ensured, in particular;

- The appropriate and approved material is used.
- The appropriate pump is used.
- Skilled labour is employed for installation of pumps.

### TESTING

- The piping connections to pumps shall be pressure tested to 1.5 times the working pressures.
- Strainer shall be cleaned after initial flushing of Chilled water piping system.
- Insulation of piping to be done after pressure testing.
- Electrical circuits / controls and connections are to be checked.

### INSPECTION

- Inspection request (IR) shall be raised for consultant's inspection.
- QC inspection shall be carried out as per the installation checklist and manufacture's instructions.
- Inspection shall be recorded in the approved format.

## SAFETY

- Work will commence as per safety regulations laid down in the contract specification and project safety plan.
- Proper safety harness to be used and secured, if required.
- All personal protective equipment shall be used as appropriate according to the nature of job.
- Housekeeping shall be of good standard and all cut lengths and debris shall be removed.
- All lifting operations shall be monitored by Safety Officer.

## Code No. : MS-AC-07

# Method Statement
# Installation of Air Handling Unit (AHU's)

| PLANNING DEPARTMENT | |
|---|---|
| Rev.: 0 | |
| Date of Issue: Feb 2007 | **MEP PLANNING MANUAL** |
| Doc No.: MEP-01-07 | |

# METHOD STATEMENT FOR INSTALLATION OF AIR HANDLING UNITS (AHU's)

## SCOPE AND PURPOSE

This "Method Statement" covers the on site installation of AIR HANDLING UNITS (AHU's) and the requirements of checks to be carried out.

## REFERENCE DOCUMENTS

- Project Specifications
- Approved shop drawings (latest revision)
- Approved material submittals

## GENERAL

Air Handling Units generally be installed in locations shown in drawings, serving corridors and supplying treated fresh air to the lobbies and apartments located at various floors.

## EQUIPMENT

- Drilling Machine
- Spirit Level
- Lifting Equipments (Cranes)
- Rollers
- Scaffolding
- Hand tools of Trades men

## RESPONSIBLE PERSONNEL

- Project Engineers
- Construction In-Charge.
- Site Engineer/Site supervisor.
- QA/QC Inspectors.
- Safety officer
- Site Foremen.
- Pipe fitters
- Helpers

## METHOD OF PRE INSTALLATION

- Ensure that approved material required to carryout work will be available.
- Check the name plate details of Air handling Units as per approved shop, drawing/schedules and material submittals before installation.
- Prior to commencement of work, area and access will be inspected to confirm that the site is ready to commence the work

| PLANNING DEPARTMENT | |
| --- | --- |
| Rev.: 0 | |
| Date of Issue: Feb 2007 | **MEP PLANNING MANUAL** |
| Doc No.: MEP-01-07 | |

- All relevant documentation and material applicable to particular section of works will be checked by site engineer before commencement.
- Physical verification of material will be carried out for any damages prior to taking from stores.
- The site engineer / supervisor will give necessary instruction to tradesman and provide necessary construction / shop drawings.
- The site engineer / supervisor will also check that tools and equipments available are in compliance to contract requirements.
- The site supervisor also explains tradesman regarding safety pre-cautions to be observed.
- Ensure that all lifting operations are carried out as per approved procedure.

## METHOD OF INSTALLATION

6. Prior to commencement of work coordination will be done with other services.

7. Check the foundation of AHU for size, orientation and finishes as approved shop drawing

8. Make sure that AMU's are free from damage and all internal components are complete and in good condition.

9. AHU's will be installed in location as per approved shop drawing.

10. Install the AHD's on concrete foundation by using suitable equipment recommended by manufacturer (eg. Rollers/jacks etc.). Manufacturer's recommendations will be followed during erection of AHU.

11. Multi-section units will be joined as per Manufacturer's instructions. Ensure to remove Shipping Bolts.

12. Ensure the orientation of AHD's as per approved shop drawing during installation.

13. Serrated rubber pads (Neoprene Isolator) will be provided below AHUs.

14. AHU's which are ceiling suspended will be mounted by using threaded rods, spring hangers etc. as per approval.

15. Any cut edges of angles, channels or threaded rods will be touch up with Zinc rich paint for ceiling suspended AHD's.

16. Connect all piping and accessories to AHU's as per approved shop drawing.

17. Ensure that Dielectric unions/Flanges are used for piping connection to AHU's.

18. Ensure that drain connections are made with adequate slope and approved U-trap.

19. Complete the duct connections to Air handling Units as shown in approved shop drawing.

20. Filters of specified sizes will be provided.

21. Check that adequate space for maintenance of Air Handling Units is provided as per approved shop drawing.

22. Complete the Electrical power connection including earthing in all respects as per approved electrical drawing and Manufacturer's recommendations.

23. Complete the BMS interfacing with DDC controllers as per approved BMS 9-drawings.

| | |
|---|---|
| PLANNING DEPARTMENT | |
| Rev.: 0 | **MEP PLANNING MANUAL** |
| Date of Issue: Feb 2007 | |
| Doc No.: MEP-01-07 | |

## QUALITY CONTROL

- QCE along with Project Engineer and site Supervisor will monitor that all components are installed as per the contract specifications and approved submittals.
- Inspection Request (IR) shall be submitted to the Main Contractor/Consultant.

## SAFETY

- Work will commence as per safety regulations laid down in the contract specification and project safety plan.
- Proper safety harness to be used and secured, if required.
- All personal protective equipment shall be used as appropriate according to the nature of job.
- Housekeeping shall be of good standard and all cut lengths and debris shall be removed.
- All lifting operations shall be monitored by Safety Officer.

# Code No. : MS-AC-08

# Method Statement
# HVAC Testing, Adjusting & Balancing

# METHOD STATEMENT FOR TESTING, ADJUSTING & BALANCING OF HVAC SYSTEM

## SCOPE AND PURPOSE

This "Method Statement" covers the on site Testing, Adjusting and Balancing of HVAC systems installed. This Method statement to be read in conjunction with project commissioning plan.

## REFERENCE DOCUMENTS

- Project Specifications
- Approved shop drawings (latest revision)
- Approved material submittals

## GENERAL

The HVAC system testing, adjusting, and balancing is the process of checking and adjusting all environmental systems in a building to produce the design objectives. This process Includes:-

- Balancing air and water distribution systems
- Adjusting the total system to provide design quantities.
- Electrical measurements.
- Establishing quantitative performance of all equipment.
- Sound measurements.

The above procedures are accomplished by

- Checking the installations for conformity to design.
- Measuring and establishing the fluid ' quantities of the system as required to meet design specifications.
- Recording and reporting the results

## EQUIPMENT

- Balometer
- Anemometer
- Micromanometer
- Pitot Tubes
- Tachometer
- Sound Meter Tester
- Water gauges (duel/single)
- Multimeter/ Tong Tester
- Hand tools of Trades men

## RESPONSIBLE PERSONNEL

- Commissioning Manager
- Seder (Specialist commissioning agency) personnel

- Equipment/ system supplier (as applicable)
- Project Engineers
- Construction In-Charge.
- Site Engineer/Site supervisor.
- QA/QC Inspectors.
- Safety officer
- Site Foremen.
- Electrician
- Helpers

## METHOD OF PRE INSTALLATION

- Ensure that approved material required to carryout work will be available.
- Ensure that As-Built marked up drawings are available reflecting the actual site installations.
- Prior to commencement of work, area and access will be inspected to confirm that the site is ready to commence the work
- Commissioning Manager along with representatives of Sender will carryout Visual inspection of all installations in order ensure that installations are in accordance with approved documents? In case of any discrepancy observed the same shall be brought to the notice of the Project engineer and Consultant for resolution of same.
- Ensure that all instruments are calibrated and are in proper working condition.
  Prior to start commissioning that it will be ensured that all chilled water pipe work is hydrostatically pressure tested satisfactorily and all documentation is available.
- Prior to Commissioning of chilled water system, it will be ensured that all piping is satisfactorily flushed chemically cleaned. And all the pipe work is reinstated to original positions.
- Accessibility of all valves, VCD's will be ensured before start of commissioning. Where required temporary access shall be provided for VCd's.

## METHOD OF INSTALLATION

1. Testing, adjusting and balancing of HVAC systems installed will be CM/SP/PE carried out as per the specialist commissioning procedures attached by specialist commissioning agency.
2. Commissioning Manager ensures that all manpower deployed by Specialist agency is competitive and sufficient to complete the TAB work.
3. All the test documentation will be recorded on the test sheets attached to the Specialist commissioning procedures.
4. Manufacturer's Representative will be associated for start equipment and as required.
5. Status of Testing and commissioning will be maintained and reported periodically to Project Managers. Consultant.

## QUALITY CONTROL

- QCE in coordination with Commissioning Manager and Project CM//PE/QCE Engineer notify to Consultants for testing and commissioning of HVAC systems area wise for witnessing the same.

- QCE in association with Commissioning Manager and Project Engineer will ensure that all the test documentation is complete and signed off.

## SAFETY

- Work will commence as per safety regulations laid down in the contract specification and project safety plan.
- All personal protective equipment shall be used as appropriate according to the nature of job.
- Housekeeping shall be of good standard and all cut lengths and debris shall be removed.

## ATTACHMENTS:-

**SEDER HVAC TAB METHOD STSTEMENT ALONG WITH REPORT FORMS.**

## Code No. : MS-AC-09

# Method Statement
# Flushing & Chemical Treatment Chilled Water System

| PLANNING DEPARTMENT | |
| --- | --- |
| Rev.: 0 | |
| Date of Issue: Feb 2007 | **MEP PLANNING MANUAL** |
| Doc No.: MEP-01-07 | |

# METHOD STATEMENT FOR FLUSHING & CHEMICAL TREATMENT CHILLED WATER SYSTEM

## SCOPE AND PURPOSE

This "Method Statement" covers the on site flushing and chemical cleaning of the Chilled Water System installed in JBR-Sector4 including towers and podium should be read in conjunction with approved material submittal

## REFERENCE DOCUMENTS

- Project Specifications
- Approved shop drawings (latest revision)
- Approved material submittals

## GENERAL

Chilled water system inc des pipes, fittings and valves used for transportation of chilled water to FCU's , AHU's through chilled water pumps. Flushing of system ensure removal of all contamination that may occur during manufacturing, storage and installation of piping. Chemical cleaning ensures the removal of oxides, oils and greases by using chemicals such as acids, alkalis, complexing agents etc;. Continuous monitoring of the condition of the cleaning solution by the cleaning specialist will be necessary throughout the chemical cleaning process.

## EQUIPMENT

- Chilled Water Pumps
- Temporary hoses for filling/ drainage
- Temporary water tanks for disposal of chemical/ flushing water

## RESPONSIBLE PERSONNEL

- AHEE MEP Coordinator
- Sr. Project Endineer HVAC
- Construction In-Charge.
- Site Engineer/Site supervisor.
- Personnel from specialist supplier
- QA/QC Inspectors.
- Safety officer
- Site Foremen.
- Electrician
- Helpers

## PROCEDURE:

## PRE REQUISITES FOR FLUSHING:

1. Chemical cleaning of the system will not be undertaken until the system installation has been completed, pressure tested and approved by consultant, vented and filled with clean water, static and dynamic flushing completed and circulating (system) pumps are available for operation.

2. Full access to all parts of system to be available, including access panels in ceilings, to enable access to valves, drains and vents etc.

3. Flushing will be carried out for each tower & each chilled water circuit (secondary and tertiary) separately. A detailed procedure is indicated here, which is common for all the circuits. Marked up schematic drawings for each secondary and tertiary chilled water circuit for each tower, for flushing are enclosed for reference. Following points are identified and marked in the enclosed schematic drawings for each circuit:

   Note: At present only one marked up schematic drawing for CO 2 TO 2 tower, secondary circuit is attached for review & comments.

   a) Number of floors covered in the circuit.

   b) Number of Terminal units (Heat exchangers, AHUs, FCUs) at various levels.

   c) Fresh water filling point.

   d) Drain point from where the system water will be drained out.

   e) Number of terminal units at various levels, where temporary bypass loops shall be provided.

4. Before flushing work commences ensure that:
   a) All pipe ends are looped/ closed prior to filling the water.

   b) All valves except drain valves (gate/butterfly/globe/DRV/commissioning set valves) are fully open. Drain valves will be closed. All 2 way control valves are fully isolated. Inlet and outlet valves for terminal units, where bypass loops are not provided will be closed to avoid flushing water passing through the coils. Water will be drained from these units at the final stage of flushing before adding final chemical.

   c) All necessary temporary bypasses installed around terminal units (heat exchangers, AHU's and FCU's) to be verified as fully open.

d)   Sufficient fresh and clean water supply is available continuously with adequate pressure at water filling point.

    e)   Drain points are properly connected / kept ready for connection to nearest drain point. Drain points will be decided in consultation with plumbing / drainage department and will be approved by Consultants.

    f)   Portable storage tanks to store drained water will be arranged, as required.

5.   The objective of the flushing and cleaning treatment process is to provide acceptable water quality (defined in 6.0) and internal pipe conditions that will permit the commissioning of the systems and provide a foundation from which an effective ongoing regime of water treatment and system management can commence.

6.   At all stages of the flushing and cleaning process the system will be offered for witnessing by the Consultant. The witnessing will include water quality, water quality, water analysis, pressure readings and strainer, deposits. The witnessing will be a continuous exercise and the full involvement of all witnessing parties will be required. During every stage of flushing process, for each circuit, samples of water will be taken from top & bottom level of each stage and tested an after satisfactory report next sage of flushing will commence.

## FLUSHING PROCESS:

Generally flushing will be carried out in following steps for each secondary an l d tertiary chilled water circuit.

### A. STATIC FLUSHING:

1.   Complete system to be filled with clean water from bottom most point of the system AND vented through air vents at various locations and at the high points of the system

2.   System pumps to be operated for a period of 1.2 hours to agitate any debris within the system

3.   With pumps switched 'off, drain system at lowest point(s) of system all drains and air vents to be opened at same time to expedite the flushing. It is to be ensured that drain point ~is,,b&40d maximum to dram any: large sized debris in side the system. Ensure that all system is &aiii6d & no, water is remained inside the pipes

4.   Check quality of drained water. Water be drained shall be collected in a separate tank (as required) and then disposed off safely or will be drained in nearest manhole available.

5.   Upon completion of draining, system to be re-filled and vented with cleans water

### B. DYNAMIC FLUSHING

6. Once the entire system' is full of water, then valves at the branch connection to the riser at each level/floor will be closed & only valves of top five levels/floors in the circuit and the plant room floor will be opened for water circulation

   The flushing velocities will be achieved by utilizing the installed circuit pumps. The number of pumps running will depend on the amount of the circuit being flushed. Prior to starting the pumps, direction, alignment and installation will be checked.

   The removal of general contaminant from the system will be achieved by using a dynamic balanced flushing procedure.

   System pumps will run and draw water to drain, but only at the rate at which clean water is introduced. At all times the system pressure will be maintained at such a level so as to exceed the static head. The system will not be drained during the dynamic flushing, to prevent the induction of air, which could accelerate the rate of corrosion, and also lead to air locks.

   During this process, water will be drained from the lowest point at the rate fresh water. is introduced in to the system. Also during this process, all the strainers in each level/floor will be opened & cleaned. Also water will be drained from the end points to ensure that dirt is not settled in the system. After every 8 to 10 hours, valves at the branch connection from the riser of the top five levels/floors will be closed & valves of next five (lower side) levels/floors will be opened for flushing. Same procedure will be followed till all the levels/floors for the entire circuit are completed. After that finally, valves at the branch connection from the riser for all the levels/floors, for the entire circuit will be kept open for water circulation for 3 to.5 hours. During this process, all the strainers in each level/floor will be once again opened & cleaned as required. At all the time during the entire process, water samples from top level & bottom level of each stage in the system will be checked for water quality.

7. The aim of the flushing process will be to initially remove all large debris from the pipe work. The pumps will be continually rotated during this process so the strainers can be cleaned. The flow in the pipes will be measured (through commissioning sets) and accordingly velocities will be recorded. These will be monitored and the strainers shall be cleaned if flow decreases. The main isolating valves may be set to create a balanced flush.

8. All strainer screens within the system will be checked and cleaned at regular intervals during the flushing process, until the screens no longer show any signs of contaminant.

9. Drain point will be kept as near to the filling point as possible to ensure that water is circulated completely in the system. Water will be drained (and make up water is introduced at the same rate), until it is as clean as the make up water. Dynamic Flushing will continue until water quality is at acceptable levels (defined in 6.0),

## CHEMICAL CLEANING:

1. Cleaning chemical M-235 will be introduced into the system. Refer to the attached table. M235 to be remained in system for 24 hours and not to exceed 72 hours.

2. The chemical will be circulated throughout the system to establish an even concentration. Water samples will be taken at the lowest point, highest point and intermediate floors (if necessary) and far points of the system (locations will be identified in chilled water riser diagram and to be approved prior to proceed with flushing) and tested at regular intervals.

3. The dynamic flushing process will then be carried out again for the entire circuit, until it is free of contaminants. The parameter for acceptance of the system will be:
   - TDS (total dissolved solids) - within 10% of incoming mains water.
   - Iron - below 1 ppm.
   - Visual - clear, bright and free from particulate matter.

4. Upon successful completion of the cleaning and flushing process, system to be re- filled with clean water and vented completely. Long term corrosion inhibitors (M-381) and a chemical biocide (M-403) will be introduced into the system, in proportion to the system volume. Refer to the attached table for each building.

5. The corrosion inhibitor and biocide will be circulated throughout the system and tests undertaken to establish full circulation and correct concentration.

6. All temporary bypass loops will be removed and all Heat exchangers, AHUs & FCU coils will be back flushed. Heat exchangers, FCU / AHU coil connections will be completed for normal working conditions.

## ACCEPTANCE CRITERIA

The acceptance criteria for the results of the witnessed measurements carried out in Clause 5.2.4 &5.3.3. are that:

### Static Flushing

(1) The TDS levels achieved at the end of each high velocity flush are not greater than 10% above the incoming mains water.

### Dynamic Flushing

a) Before Cleaning Chemical Addition -
Water appearance                          : Clear, Yellowish.

Iron Level      : Less than 10ppm.

TDS        : within 10% of incoming mains water.

b) After Cleaning Chemical Addition - Water

appearance     : Clear.

Iron Level      : Less than 1 ppm.

TDS        : within 10% of incoming mains water.

(2) The velocities during each high velocity flushing stage are in excess of 1.36 m/s in the largest pipe in the system

## DISPOSAL OF EFFLUENT:

The chemically treated water shall be discharged in to mobile tankers and discharged properly as per local authority regulations.

## QUALITY CONTROL

- Project Engineer along with Specialist supplier will monitor the flushing process and ensures that same is in full compliance as per approved submittals.

- Quality Control Engineer will coordinate with Consultant for all witnessing activities during chemical cleaning process

## SAFETY

- Where chemical cleaning is being carried adequate notices will be displayed for " NO SMOKING" and Warning Signs.
- All tradesmen engaged will be given proper safety orientation for chemicals being used. And appropriate personnel protective equipments will be provided.
- Continuous monitoring will be carried out by trained operatives for any emergency actions.
- Ensure that good house keeping at all times.

# Section 11

# Method Statement for Plumbing Installation

## Code No. : MS-PL-01

# Method Statement
# Concealed Drainage Piping

| PLANNING DEPARTMENT | |
| --- | --- |
| Rev.: 0 | |
| Date of Issue: Feb 2007 | **MEP PLANNING MANUAL** |
| Doc No.: MEP-01-07 | |

# METHOD STATEMENT FOR CONCEALED DRAINAGE PIPING

## SCOPE AND PURPOSE

This "Method Statement" provides the detailed account of the installation and testing of concealed drainage pipes in toilets from Level 1 and up in all towers.

## REFERENCE DOCUMENTS

- Project Specifications
- Approved shop drawings (latest revision)
- Approved material submittals

## GENERAL

Installation of drainage pipes within the concrete slab in toilets, garbage rooms, balconies etc from level 1 and up in towers, shall be done in accordance with the project specifications and drawings.

## EQUIPMENT

- Spirit Level, Water Level
- Measuring tape, Nylon twine
- Pipe cutter, Hack saw, Files, Trimmers etc
- Marking tools, Plumbers tool box, Hammers

## RESPONSIBLE PERSONNEL

- Project Engineers
- Site Engineer/Site supervisor
- QA/QC Inspectors
- Safety officer
- Site Foremen
- Plumbers
- Helpers

## METHOD OF PRE INSTALLATION

1. Approved materials shall be purchased in accordance with the approved shop PE / SS drawings and For Construction Contract Specifications.

2. Proper materials including pipes and fittings are to be selected and used PE / SS according to approved shop drawing and good engineering practices.

3. Adequate stock of material, tools and tackles, testing equipments and PE / SS consumables along with sufficient man power shall be arranged to carry out the work.

4. Site supervisors are to be provided with the civil work schedules for the following works:

   a. Shuttering works
   b. Bottom Reinforcement fixing works
   c. Top Reinforcement fixing works and
   d. Concreting works

5. Site supervisors are to be provided with the finished floor level as well as structural floor level by the civil surveyors at site before starting marking of pipe line route on shuttering.

6. Ensure that the location of masonry walls, concrete walls, 'shafts, grid lines etc are marked by the main civil contractor on slab shuttering.

7. Thickness of plastering, tiling etc with tile modules and thickness of screed and flooring are to be finalized and produced at site before execution of piping work.

8. Ensure that the construction power and water for testing of pipes are provided by the main contactor near to the work place

9. Safe and tidy work place, clear access to the work place, modes to shift material, tools and tackles etc are to be ensured.

10. Civil works including reinforcement schedules should be designed to accommodate the drainage pipes with proper slope and level.

**PIPING :**

1. Main Contractor's Surveyor to mark the locations of Floor gullies, sanitary fittings etc as per approved shop drawing and revised architectural drawing on shuttering before putting the reinforcement. Information regarding completion of marking to be given to VL Supervisors etc.

2. Main Contractor to fix bottom reinforcement and inform VL supervisors. VL to install all pipes and fittings as per approved shop drawing and as follows.

   1. Ensure that the pipes are cut square before they are placed inside the reinforcement.
   2. Use cleaning fluid on both surfaces to be joined. This removes all dirt and machine

| PLANNING DEPARTMENT |
| Rev.: 0 |
| Date of Issue: Feb 2007 |
| Doc No.: MEP-01-07 |

**MEP PLANNING MANUAL**

ARAB MEP
www.arabmep.com

release agents for the chemical solvent weld. Failure to do this can result in joint failure.

3. Apply solvent cement evenly over mating surfaces of both pipe and socket.
4. Insert pipe into socket with slight twisting action to full socket depth.
5. Surplus cement should be removed with a cloth.
6. The joint should be firm enough to handle in 5 minutes.
7. Pipe runs are to be placed and tied to reinforcement considering proper slope and gradient as per approved shop drawings.

3. Pipes and fittings which are push fit types are to be connected as follows

   1. Lubricate the pipes cut in square and chamfered as well as fittings with rubber lubricant and push fit to full socket depth.
   2. Withdraw pipe 5 mm on waste system and 10 mm on soil system to allow for expansion.
   3. Soil fittings with spigot ends should be inserted into sockets to depth marks engraved on spigot.
   4. This procedure automatically allows for expansion.
   5. Pipe runs are to be placed and tied to reinforcement considering proper slope and gradient as per approved shop drawings.

4. Main Contractor to install Top reinforcement in the slab taking adequate care to ensure that drainage pipes / fittings are not damaged / disturbed etc. VL to ensure that pipes are tied securely to bottom, top reinforcement and stirrups so that they are not displaced at the time of concreting.

5. All Fire Stopping related works to MEP services viz. pipes entering / exiting Fire rated Walls etc to be done by Main Contractor.

## TESTING :

1. Hydraulic testing of the pipes jointed with solvent .cement should be done after 1 hours of jointing.

2. Check for gradient and slope before filling water.

3. Extend one of the pipe ends to attain the required head of 1.5 m for hydraulic testing.

4. Compression type rubber plugs are` to be used to plug all open pipe ends.

5. Fill the entire pipe network with fresh clean water up to the required height of 1.5 m from the highest point.

6. Keep the water with the head required for 6 hours for testing.

7. Check the joints for any leakage also the fall in water level in the Vertical Pipes. Any leakage found is to be rectified immediately and tested again.

8. Concreting to be done only after hydraulic testing of pipes has been successful and witnessed by the engineer.

9. Concreting to be poured over the pipe using mortar pans and compacting to such that needles do not touch the hydraulically tested pipes. Main Contractor to ensure that the pipes / fittings etc are not disturbed during concreting operations. This is critical to ensure alignment of pipe work is maintained as well as preventing damage.

10. Water filled into pipes to be retained and monitored for the entire duration of concreting so that at any displacement of joints and, subsequent leakages can be monitored.

11. If the above mentioned situations arise, concreting over that particular area is to be stopped, 'and rectification of displaced, pipes,' sealing of joints etc to be taken up immediately. Main contractor to wait until the rectification is carried out properly to VL Supervisor/Foreman's satisfaction.

12. Follow items 1-8 above after rectification.

## POST INSTALLATION PROCEDURE

1. Ensure to prepare As-Built Drawings as soon as possible after the work is completed.

2. Location of the concealed drainage piping will be marked on the suffix as per As-built or redline (As-built marked up) drawings to avoid any drilling or related activity to prevent damage to the concealed drainage piping.

## QUALITY CONTROL

- QCE along with Project Engineer and site Supervisor will monitor that all components are installed as per the contract specifications and approved submittals.

- Inspection Request (IR) shall be submitted to the Main Contractor/Consultant during the following stages of work have been completed :-
  - Inspection of Piping installation and hydraulic testing

## SAFETY

- Work will commence as per safety regulations laid down in the contract specification and project safety plan.
- Proper safety harness to be used and secured, if required.
- All personal protective equipment shall be used as appropriate according to the nature of job.
- Housekeeping shall be of good standard and all cut lengths and debris shall be removed.

<u>Code No. : MS-PL-02</u>

# <u>Method Statement</u>
# <u>Installation of Above Ground Drainage Piping</u>

| PLANNING DEPARTMENT | |
| --- | --- |
| Rev.: 0 | |
| Date of Issue: Feb 2007 | **MEP PLANNING MANUAL** |
| Doc No.: MEP-01-07 | |

# METHOD STATEMENT FOR INSTALLATION OF ABOVE GROUND DRAINAGE PIPING

## SCOPE AND PURPOSE

This "Method Statement" covers the on site installation of above ground drainage piping and the requirements of checks to be carried out.

## REFERENCE DOCUMENTS

- Project Specifications
- Approved shop drawings (latest revision)
- Approved material submittals

## GENERAL

Above Ground drainage piping is generally be installed in podium, risers and lobby areas as per locations shown in approved shop drawings.

## EQUIPMENT

- Drilling Machine
- Spirit Level
- Measuring Tape
- Nylon twine
- Pipe Cutter, Hack Saw, Files Trimmers etc
- Hammers
- Ladders
- Scaffolding
- Hand tools of Tradesmen

## RESPONSIBLE PERSONNEL

- Project Engineer (Plumbing)
- MEP Coordinator(Main Contractor)
- Construction In-Charge
- Site Engineer/Site supervisor
- QA/QC Engineer
- Safety Officer
- Site Foremen
- Helpers

## METHOD OF PRE INSTALLATION

- Ensure that Above Ground Drainage Piping Material received are inspected and found acceptable as per approved material submittal, are available to carry out the work.
- Proper Material including pipes and fittings are to be selected and used in accordance with approved shop drawings.
- Ensure that all related material (Supports/Solvent Cement/Lubricants) of approved make is available before carrying out any work.
- Prior to Commencement of work, areas and access will be inspected to confirm that site is ready to commence the work and all relevant shop drawings dully approved for construction.
- All relevant documentation (Drawings) and Material applicable to particular section of works will be checked by Site Supervisor prior to commencement of work.
- The Site Engineer/ Site Supervisor will give necessary instructions to tradesman (Plumber) and provide necessary approved construction/ shop drawings to latest revision along with coordinated lay outs.
- Ensure that location of grid lines, reference levels are marked by Main Contractor.

- *INSULATION OF HORIZONTAL RUNS*

  3.1     Mark the route of above ground drainage piping on the soffit of slab.

  3.2     Determine the position of support and fix the supports using anchor bolts and ensure that all fixings (threaded rods, angles, clamps etc.) are straight and secure.

  3.3     Install the Drainage pipes on to the supports. Ensure proper slope and gradient is maintained for all horizontal runs of Drainage piping.

  3.4     Install all pipes and fittings as per the following procedure .

  - Ensure that the pipes are cut square before they are joined.

  - Use cleaning fluid on both surfaces to be joined. This removes all dirt and machine release agents for chemical solvent weld. Failure to do this can result joint failure.

  - Apply solvent cement evenly on mating surfaces of both pipes and socket.

  - Insert pipe into socket with slight twisting action and full socket depth.

  - Surplus cement should be removed with a cloth.

  - The joint should be firm enough to handle in 5 minutes.

| PLANNING DEPARTMENT | |
|---|---|
| Rev.: 0 | |
| Date of Issue: Feb 2007 | **MEP PLANNING MANUAL** |
| Doc No.: MEP-01-07 | |

3.5     Pipes and fittings which are push fit types are to be connected as follows:

- Lubricate the pipes cut in square and chamfered as well as fittings with rubber lubricant and push fit to full socket depth.

- With draw pipe 5mm on waste system and 10mm on soil system to allow it for expansion.

- Soil fittings with spigot ends should be inserted into sockets to depth marks engraved on spigots. This procedure automatically allows for expansion.

- ### *INSULATION OF VERTICAL RISERS*

3.1     Ensure that the shafts are clear and through up to maximum 6 floors above.

3.2     Ensure that the required provision for maintenance of shaft piping is provided in the block work as per approved drawing.

3.3     The inside faces of shaft to be finished before fixing of the vertical riser pipes.

3.4     Mark sure that proper working platforms are made for safe working inside the shaft.

3.5     Make sure that all pipes protruding out from the slab of the corresponding toilets are in plumb.

3.6     Mark the location of vertical riser and fix the supports (as per approval) at intervals as specified.

3.7     Fix the pipe fittings on the outlet pipes protruding from the slab and align to the plumb of the vertical riser.

3.8     Joining of pipes and fittings will be carried out as per clause 2.4 & 2.5 above.

3.9     Fix the clamps (as per approval) to vertical riser pipes at specified intervals of 1.6M which are already fixed to the structural members.

3.10   At every five (5) levels additional support with U- Clamp will be provided. Anchor supports to be provided for vertical loads at each directional change.

- Each directional change or water testing of riser pipes will be carried out after completing the installation of riser pipe.
- Riser to be tested with minimum 1.5m water head at eight point / air tested. While testing risers all horizontal floor toilets / WP's etc. will be disconnected / isolated.

| | |
|---|---|
| PLANNING DEPARTMENT | |
| Rev.: 0 | **MEP PLANNING MANUAL** |
| Date of Issue: Feb 2007 | |
| Doc No.: MEP-01-07 | |

## QUALITY CONTROL

- QCE along with Project Engineer and site Supervisor will monitor that all components are installed as per the contract specifications and approved submittals.

- Inspection Request (IR) shall be submitted to the Main Contractor/Consultant after completion of installation of riser pipes along with testing.

## SAFETY

- Work will commence as per safety regulations laid down in the contract specification and project safety plan.
- Proper safety harness to be used and secured, if required.
- All personal protective equipment shall be used as appropriate according to the nature of job.
- Housekeeping shall be of good standard and all cut lengths and debris shall be removed.
- Ensure that proper scaffolding/ ladders are available.
- Ensure that shafts are adequately blocked above and below the working level, so that no fall of debris/ materials takes place.

## Code No. : MS-PL-03

# Method Statement
# Domestic Water (Hot & Cold) Supply Piping

| PLANNING DEPARTMENT | |
| --- | --- |
| Rev.: 0 | |
| Date of Issue: Feb 2007 | **MEP PLANNING MANUAL** |
| Doc No.: MEP-01-07 | |

# METHOD STATEMENT FOR DOMESTIC WATER (HOT & COLD) SUPPLY PIPING

## SCOPE AND PURPOSE

This "Method Statement" covers the on site installation, testing and insulation of the Domestic Water Supply Piping including Secondary risers and the requirements of checks to be carried out.

## REFERENCE DOCUMENTS

- Project Specifications
- Approved shop drawings (latest revision)
- Approved material submittals

## GENERAL

Domestic Water Supply Piping System includes Hot Water Supply Piping, Cold Water Supply Piping, Risers, fittings, valves and accessories used for supply of water to Toilets, Kitchen and Utility rooms.

## EQUIPMENT

- Drilling Machine
- Cutter for PVC Pipes
- Welding machine for PP-R
- Pressure Test Pump
- Hammers
- Ladders
- Scaffolding
- Hand tools of Tradesmen

## RESPONSIBLE PERSONNEL

- Project Engineer (Plumbing)
- MEP Coordinator(Main Contractor)
- Construction In-Charge
- Site Engineer/Site supervisor
- QA/QC Engineer
- Safety Officer
- Site Foremen
- Helpers

## METHOD OF PRE INSTALLATION

1. Ensure that approved materials are available to carry out the work.

2. Proper materials including pipes, fittings and associated accessories are with drawn from stores according to approved shop drawing and good engineering practices.

3. Physical verification of materials will be carried out for any damages prior to taking from stores.

4. Prior to Commencement of work , areas and access will be inspected to confirm that Site is ready to commence the work.

5. All relevant documentation (Drawings) and Material applicable to particular section of works will be checked by Site Supervisor prior to commencement of work.

6. The Site Engineer/ Site supervisor will give necessary instructions to tradesmen (Pipe fitters/welders) and provide necessary approved Construction/Shop Drawings along with coordinated layouts.

7. The Site Engineer/Foremen will also check that proper tools and equipment are available to carry out the work and are in compliance with contract specification.

8. The Site Engineer also explains the tradesmen regarding safety precautions be observed.

9. Prior to Hydrostatic Pressure testing, Site Engineer will ensure that calibrated pressure gauges are available and are in good condition.

## METHOD OF INSTALLATION

1. Foremen will carryout a site survey and mark the route of water supply Piping ( Hot & Cold) as per approved shop drawings. In the event that there are any discrepancies or difficulties in executing the work, these will be brought to the notice of Project Engineer for corrective action.

2. Co-ordination with other trades will be carried out while marking the route of piping.

3. Determine the position of support and fix the supports using anchor bolts and ensure all fixing are tight and secure. While drilling the soffits, ensure that there is no damage to embedded services in the slab.

4. Any cut edges of angles, channels or threaded rods will be touch up with Zinc rich paint.

5. Install the pipes in position on the supports. Wherever possible pre-assembly of pipes and fittings at ground level will be carried out.

6. Assemble the pipes and fittings as per approved shop drawing.

7. Assemble of U-PVC pipes and Fittings by using Solvent cement. Ensure that joints are clean before applying solvent cement.

8. For joining of PP-R Pipes and Fittings manufacturer's recommendations will be followed (Copy enclosed).

9. For fixing PEX pipes, ensure that wall chases are done properly by Main Contractor as per locations shown in approved shop drawing.

10. Fix the female elbow with cover to be located at proper height as per approved shop drawing. Ensure the line and alignment with wall tiling.

11. Install the PEX pipes with sleeves (Conduit) in the wall chases. Ensure that joint between PEX pipe and female elbow is proper and leak free.

12. Fix the PEX pipe to Water supply lines by using male adaptors.

13. Spacing between supports / hangers will be maintained in accordance with latest approved shop drawings/ manufacturer's recommendations

14. Ensure all open ends of pipes, fittings and valves are covered with polyethylene sheet before leaving work space.

## 15. INSTALLATION OF VALVES AND ACCESSORIES

15.1 Install system valves and accessories as per latest approved shop drawings.

15.2 Ensure that system equipment, valves and accessories are secure and rigid.

15.3 The installation shall be done allowing sufficient access to all valves/ strainers/ gauges as per Manufacturer's recommendations.

## 16 INSTALLATION OF WATER SUPPLY RISERS

16.1 Pipe sizes will be identified first as per latest approved shop drawing and shifted to respective floors.

16.2 Install the supports as per approved Shop drawing/ Manufacturer's recommendations.

16.3 Joining of pipes will be carried out as per above procedure (refer clause 6.2.7 & 8).

16.4 After installation of risers check the pipeline for proper alignment and supports.

## 17 HUDROSTSTIC PRESSURE TESTING

17.1 Complete pipe work will be subjected to hydraulic pressure tested as per technical specification. Depending on ongoing Construction activities sectional hydro testing will be under taken to meet the requirements of the programmed. Test pressure will not be less than 1.5 times the working pressures but not less than 1035 KPa (for Two hour period) which ever is greater. Prior to any testing the system pressure will be shown on the pressure testing documentation.

| PLANNING DEPARTMENT | |
| --- | --- |
| Rev.: 0 | |
| Date of Issue: Feb 2007 | **MEP PLANNING MANUAL** |
| Doc No.: MEP-01-07 | |

17.2 Pressure gauges used for Pressure testing will have valid calibration certificate.

17.3 After successful Pressure testing ensures that piping system is fully drained and released for insulation and other related activities.

## 18 INSULATION

18.1 Insulation of water pipe work will be carried out as per details as shown in approved submittals. Thermal insulation of joints will be carried out after pressure testing.

18.2 Make sure that Pipes are clean before applying any insulation. Follow manufacturer's recommendation for insulation of water supply piping (copy enclosed).

18.3 Ensure thickness of insulation is as per approved drawing.

18.4 Identification Bands/ Labeling will be located at is at access panel locations.

## QUALITY CONTROL

• QCE along with Project Engineer and site Supervisor will monitor that all components are installed as per the contract specifications and approved submittals.

• Inspection Request (IR) shall be submitted to the Main Contractor/Consultant for the following stages:-
   o After completion of installation prior to Hydrostatic Pressure Test.
   o Pressure testing of piping
   o After completion of insulation before ceiling clousure.

## SAFETY

• Work will commence as per safety regulations laid down in the contract specification and project safety plan.
• Proper safety harness to be used and secured, if required.
• All personal protective equipment shall be used as appropriate according to the nature of job.
• Housekeeping shall be of good standard and all cut lengths and debris shall be removed.

## Code No. : MS-PL-04

# Method Statement
# Installation of Water Heaters

# METHOD STATEMENT FOR INSTALLATION OF WATER HEATER

## SCOPE AND PURPOSE

The scope and purpose of this method statement is to define the method of storage, handling, installation and inspection of the Water Heaters as per specification and manufacturers instructions

## REFERENCE DOCUMENTS

- Project Specifications
- Approved shop drawings (latest revision)
- Approved material submittals

## EQUIPMENT

- Drilling Machine
- Cutter for PVC Pipes
- Gloves
- Spirit Level
- Measuring Tape
- Ladders
- Scaffolding
- Hand tools of Tradesmen

## RESPONSIBLE PERSONNEL

- Project Engineer
- Site Engineer/Site supervisor
- QA/QC Engineer
- Safety Officer
- Site Foremen
- Plumbers
- Helpers

## METHOD OF PRE INSTALLATION

### 4.1 Receiving

- When received at site, heaters shall be checked for quantities, model numbers and physical damages, if any discrepancies are noticed, inform supplier for replacement of the same.

- P&T Valves shall be checked for size, model and quantity.

- Check for Test Certificates, Country of Origin Certificate, Spares (if any) and Operation and Maintenance Manual for the supplied heaters.

- Request for Inspection shall be raised for Consultants inspection.

- The storekeeper, engineer and QC Engineer of ETA shall conduct receiving inspection.

- Any items found damaged or not found suitable as per the project requirements shall be quarantined, non-compliant materials shall be clearly marked and stored separately to prevent any inadvertent use until returned to vendor.

## 4.2    Storage

- Upon completion of receiving QC inspection the heaters will be segregated model / size wise and stored accordingly for easy retrieval.

- Water heaters shall be stored on a flat surface in ventilated and covered area and protected from dust.

- Inlet, outlet and return point's blanks shall not be removed until ready for connection to pipe work.

- Manufacturer's instruction shall be strictly followed as applicable.

- Storekeeper will be responsible for proper storage and maintenance of records, as required

## 4.3    Preparation

- Check and ensure that the shop drawings used are latest and approved for construction.

- Check coordination with other services prior to the installation.

- Check the piping supports locations and power supplies routing locations in coordination water heater and piping layout and ensure it does not obstruct the space around water heater for removal and maintenance.

- Ensure easy access and sufficient clearance for servicing and maintenance i.e. for replacement of water heaters, thermostat, heating elements.

## METHOD OF INSTALLATION

### 16. Ceiling Suspended Water Heaters

- Install the fabricated water heater support as per approved details.

- Install the water heater on the support.

- Remove the end caps on the inlet, outlet points. Complete the piping and valve package installation as per approved drawings. .

- Install the electrical power connections as per approved drawings.

- Check and ensure availability of adequate access for removal and maintenance of water heater.

- Provide grounding wiring as per approved drawings / suppliers instruction.

- Ensure compliance to the manufacturers instructions while installing the water heaters.

- After completion of the installation, Inspection Request shall be raised for Consultants approval.

| | |
|---|---|
| PLANNING DEPARTMENT | |
| Rev.: 0 | **MEP PLANNING MANUAL** |
| Date of Issue: Feb 2007 | |
| Doc No.: MEP-01-07 | |

### 17. Floor Mounted Water Heaters

- Mark the locations of the Water Heater base frame and hole locations on the foundation.

- Drill the suitable size holes in the foundations.

- Install the water heater on the foundation.

- Remove the end caps on the inlet, outlet points. Complete the piping and valve package installation as per approved drawings.

- Install the electrical power connections as per approved drawings.

- Connect the P&T valve to the nearest floor drain.

- Check and ensure availability of adequate access for removal and maintenance of water heater.

- Provide grounding wiring as per approved drawings / suppliers instruction.

- Ensure compliance to the manufacturers instructions while installing the water heaters.

- After completion of the installation, Inspection Request shall be raised for Consultants approval.

## QUALITY CONTROL

- QCE along with Project Engineer and site Supervisor will monitor that all components are installed as per the contract specifications and approved submittals.

- Inspection Request (IR) shall be submitted to the Main Contractor/Consultant after completion and water heater installation and pipe connection.

## SAFETY

- Work will commence as per safety regulations laid down in the contract specification and project safety plan.
- Proper safety harness to be used and secured, if required.
- All personal protective equipment shall be used as appropriate according to the nature of job.
- Housekeeping shall be of good standard and all cut lengths and debris shall be removed.

## Code No. : MS-PL-05

# Method Statement
# Installation of Domestic Water Supply Pumps

# METHOD STATEMENT FOR INSTALLATION OF DOMESTIC WATER SUPPLY PUMPS

## SCOPE AND PURPOSE

The scope and purpose of this method statement is to define the method of storage, handling, installation and inspection of the Domestic Water Supply Pumps and the controller as per specification and manufacturers instructions.

## REFERENCE DOCUMENTS

- Project Specifications
- Approved shop drawings (latest revision)
- Approved material submittals

## EQUIPMENT

- Drilling Machine
- Cutter for PVC Pipes
- Gloves
- Spirit Level
- Measuring Tape
- Fork Lift, Crane
- Scaffolding
- Hand tools of Tradesmen

## RESPONSIBLE PERSONNEL

- Project Engineer
- Site Engineer/Site supervisor
- QA/QC Engineer
- Safety Officer
- Site Foremen
- Plumbers
- Helpers

## METHOD OF PRE INSTALLATION

### 4.4 Receiving

- When received at site, each pump, valves, control panels, float switches shall be checked for quantities, Model Nos., physical damages etc. and notify supplier of any discrepancies for suitable rectification or replacement

- Check for Test Certificates, Country of Origin Certificate, Spares (if any) and Operation and Maintenance Manual for the supplied pumps.

| PLANNING DEPARTMENT | |
|---|---|
| Rev.: 0 | |
| Date of Issue: Feb 2007 | **MEP PLANNING MANUAL** |
| Doc No.: MEP-01-07 | |

- Request for Inspection shall be raised for Consultants inspection.

- Valves shall be segregated as per sizes/models and stored on racks within a covered store.

- Any items found damaged or not found suitable as per the project requirements shall be quarantined. Non-compliant material shall be clearly marked and stored separately to prevent any inadvertent use until returned to vendor.

### 4.5    Storage

- Pump shall be stored on a flat surface in well ventilated storage area.

- Inlet and outlet flange blanks shall not be removed until ready for connection to pipe work.

- Manufacturer's instructions shall be strictly followed as applicable.

- If the pumps are stored for longer periods the shaft shall be periodically rotated and lubricated, if required.
- The stored pump should be inspected periodically for obvious conditions such as' . standing water, parts theft, excess dirt buildup or any other abnormal condition.

- Storekeeper will be responsible for proper storage and maintenance of records, as required.

### 4.6    Preparation

- Civil Contractor (AHEE) shall provide the foundations designed to meet the vibration and sound control requirements.

- Check and ensure that the shop drawings used are latest and approved for construction.

- ETA shall co-ordinate the location of foundation as per approved shop drawings.

- The foundation surface shall be flat and level and smoothly finished top surface.

- Check the piping support locations and cable tray routing locations in co-ordination with pump and piping layout and ensure they are not obstructing the space around pump.

- Ensure easy access and sufficient clearance for servicing and maintenance i.e for replacement of pump, motor, pressure vessel.

## METHOD OF INSTALLATION

- Mark the locations of the pump base frame and hole locations.

- Drill the suitable size holes in the foundations.

- The pump and the other associated accessories including the piping manifold are pre-assembled on a base frame. Shift the pumps to the place of installation in safe manner. Use hand trolley / folk-lift/ crane as applicable/ required as per site conditions.

- Check and ensure free rotation of the shaft.

- Position the pump frame assembly on the foundation and fix the anchor fasteners.

- Water level the pump assembly by placing the shim plates below the base frame as required. After the installation the pump supplier shall recheck the gap between ' motor and pumps before testing as applicable.

- Position the pressure vessel and do the interconnecting pipe work as per approved drawings.

- Ensure proper coupling guards are provided if required.

- Complete the piping and valve package installation as per approved drawings. Remove the end caps fixed on the inlet flange.

- Install the electrical control panel and power connections as per approved 'r drawings.

- Provide grounding wiring as per approved drawings / manufacturers instruction.

- Follow the manufacturer's instructions while installing the pump.

- After completion of the installation, it shall be checked and certified by the local p supplier.

## QUALITY CONTROL

- QCE along with Project Engineer and site Supervisor will monitor that all components are installed as per the contract specifications and approved submittals.

- Inspection Request (IR) shall be submitted to the Main Contractor/Consultant after completion and water heater installation and pipe connection.

## SAFETY

- Work will commence as per safety regulations laid down in the contract specification and project safety plan.
- Proper safety harness to be used and secured, if required.
- All personal protective equipment shall be used as appropriate according to the nature of job.
- Housekeeping shall be of good standard and all cut lengths and debris shall be removed.

| | |
|---|---|
| PLANNING DEPARTMENT | |
| Rev.: 0 | **MEP PLANNING MANUAL** |
| Date of Issue: Feb 2007 | |
| Doc No.: MEP-01-07 | |

## Code No. : MS-PL-06

# Method Statement
# Installation of Sanitary Fixtures

| PLANNING DEPARTMENT | |
| --- | --- |
| Rev.: 0 | MEP PLANNING MANUAL |
| Date of Issue: Feb 2007 | |
| Doc No.: MEP-01-07 | |

# METHOD STATEMENT FOR INSTALLATION OF SANITARY FIXTURES

## SCOPE AND PURPOSE

This "Method Statement" covers the on site installation of sanitary fixtures, fittings and various accessories associated with sanitary ware.

## REFERENCE DOCUMENTS

- Project Specifications
- Approved shop drawings (latest revision)
- Approved material submittals

## EQUIPMENT

- Drilling Machine
- Spirit Level
- Nylon twine
- Hand tools of Tradesmen

## RESPONSIBLE PERSONNEL

- Project Engineer
- Site Engineer/Site supervisor
- QA/QC Engineer
- Safety Officer
- Site Foremen
- Plumbers
- Helpers

## METHOD OF PRE INSTALLATION

1. Ensure that all Sanitary Ware and associated accessories supplied by Main Contractor are as per approved material submittals and are inspected and found acceptable as per approved material submittals and are available to carry out the work.

2. On receiving the Sanitary Ware for installation, careful inspection of material will be done for damage to the material. Such damaged material will be returned to supplier through the Main Contractor after making a proper inspection report..

3. The Sanitary Ware will be stocked in properly ventilated stores as per the recommendations of the manufacturer. Care will be taken that no heavy material will be kept in overhead locations above the storage space for Sanitary Ware to avoid any damages. While storing the Sanitary Ware metalto ceramic contact will be avoided as.-possible. While storing sufficient spacing will be ensured for easy lifting and loading. Storage will be provided by the Main Contractor.

4. Prior to commencement of work, areas and access will be inspected to confirm that the Site is ready to commence the work.

5. All relevant documentation drawings) and Material applicable to particular section of works will be checked by Site Supervisor prior to commencement of work.

6. The Site Engineer/ Site Supervisor will give necessary instructions to tradesmen and provide necessary approved construction/shop drawings of the latest revision along with coordinated layouts.

7. The Site Engineer/Foremen will also check that proper tools and equipment are available to carry out the work and are in compliance with the contract specification.

8. The Site Engineer will also explain to the tradesmen regarding safety precautions to be observed.

## METHOD OF INSTALLATION

1. Prior to starting the installation of Sanitary fixtures, fittings and accessories it will be ensured that all pipe work in the toilets/kitchens are installed properly and tested according to approved methods, and there is no damage to the pipe work In short all first fix and second fix of the plumbing services will be completed before sanitary wares are installed.

2. All pipe supports, valves, floor drains, floor clean out will be checked for accuracy of installation.

3. The locations and sizes of piping and drainage outlets will be checked for their compatibility with sanitary fixtures.

4. All other works, especially floor and wall tile work painting are completed prior to installation of sanitary ware.

5. Ensure that all drainage pipe work is flushed out and debris removed from the pipe work.

6. Identify the models of Sanitary Ware to be installed as per approved drawings/submittals.

## 7. WATER CLOSET

- Ensure that no damage has occurred due to transport to site, and all fittings, fixing screws are available inside the packing.

- Fix the water supply angle valve to the pipe connection on the wall with escutcheon plate.

- Cut the soil drain pipe to the required length to fit the W.C. bowl outlet.

- Install the approved sealing flange connector supplied by the manufacturer on the drainage outlet.

- Place the closet in the locations and adjust to the dimensions as per
- Approved shop drawing. Follow manufacturer installation instructions.
- Mark the holes for fixing screws on the floor, and drill with correct size drill bit to the required depth. Place the anchors in the drill holes.
- Clean the floor under the W.C. closet and put in place fixing studs carefully observing the levels at all times.

- Seal the base of the water closet unit with approved sealant and allow it to dry.

- Assemble the flushing mechanism of the water cistern ensuring that all rubber seals for the screws and water connecting pipes are in place.

- Connect the cistern with water supply angle valve using a connecting piece of pipe of appropriate length supplied along with Sanitary ware.

- Fix the seat and cover to the W.C. bowl using fixing screws provided by the manufacturer.

- Secure the W.C. unit against misuse/damage by using proper protection.


## 8. <u>BIDET</u>

- Check the location and position of bidet based on approved shop drawings.

- Re-examine the rough in connections for hot and cold water supply and waste outlet trap connection.

- Fix the angle valves with escutcheon plate on the rough in pipe connections for the cold and hot water supply and tighten till flush to the wall.

- Cut the waste outlet pipe to the required length and fix the waste coupling provided on the pipe end as per manufacturer recommendations.

- Fix the mixer on the bidet and tighten the underside nut, connect the plug lever arm to the waste coupling.

- Place the bidet in position and mark the holes for the fixing screws ensuring that waste opening is aligned with waste piping.

- Drill the holes with correct size of drill bit to the required depth and anchor fasteners. On completion clean the floor in position of the bidet.

- Place the bidet in position and insert the fixing screws in the holes and tighten them carefully.

- Connect the Bidet mixer and the angle vales by using a connection piece of pipe supplied and tighten them with coupling nuts.

- Place the waste strainer in the drain opening of bidet.

- Apply sealant around the base of the bidet.

- Secure bidet against misuse/ damage by using proper protection.


## 9. <u>BATH TUB</u>

- Check the location and type of bath tub and mixer according to approved shop drawing.

- Examine the rough in piping connections for hot and cold water supply and waste out let. Check the distances and tolerances as per approved shop drawing, and manufacturer's recommendations.

- Mark the position of the bath tub and ensure the waste connection is centered with the waste coupling and the height from the floor is correct as per approved drawing.

- Place the bath tub on block work as recommended by manufacturer.

- Adjust the level of the tub using s spirit level and fix the waste strainer on the waste opening and tighten the screws.

| | |
|---|---|
| PLANNING DEPARTMENT | |
| Rev.: 0 | **MEP PLANNING MANUAL** |
| Date of Issue: Feb 2007 | |
| Doc No.: MEP-01-07 | |

- Fix the bath tub Make by placing the concrete, mortar under and around it as recommended by manufacturer. Ensure that there is no damage to the bath tub while carrying out civil works around it. Civil works by Main Contractor.

- After civil works are completed and location is clean and ready, fix the bath tub mixer and hand shower, shower rail as per approved drawing.

- Protect the bath tub against any misuse/ damage by appropriate methods.

## 10. SHOWER TRAY

- Examine the location and position of the shower tray according to the approved shop drawing.

- Check in rough in piping connections for hot and cold water supply to the shower mixer and waste piping connections for their locations according to approved shop drawing.

- Place the shower tray in position and align the waste opening. Adjust the level of shower tray using sprit level.

- Fix the waste strainer on the waste opening and tighten the screws.

- Place the concrete mortar under the shower tray according to manufacturer's instructions and complete the finishing works regarding tiling and wall finishes shower cubicles etc. as per approved shop drawing.

- Place the protective coverings over shower tray and mixer.

## 11. WASH BASIN

- Examine the location and place of wash basin as per approved shop drawing.

- Identify the model of the wash basin, check all dimensions, tolerances, and levels of the rough in water and drainage piping are matching with the wash basin to be installed as per approved shop drawing.

- Mark the holes for fixing stud on the walls, using template provided by the manufacturer.

- Drill the holes with the correct size of drill bit to the required depth and anchor fasteners in position.

- Place the threaded studs in the holes and fix the wash basins with washer and nuts.

- Check the level of the wash basin by spirit level and coordinate the levels with counter top position.

- Install bottle trap on waste out, fix the strainer on wash basin and tighten the strainer screw.

- Connect to the waste pipe on the wall and fix the escutcheon plate supplied by manufacturer to conceal piping joint.

- Fix the angle valves for cold and hot water supply flush to the wall, using escutcheon plates.

- Install water mixer on the wash basin and connect the mixer and angle valves by using tubes. Tighten the coupling nuts ensuring that installation is tidy.

| PLANNING DEPARTMENT | |
| --- | --- |
| Rev.: 0 | MEP PLANNING MANUAL |
| Date of Issue: Feb 2007 | |
| Doc No.: MEP-01-07 | |

- After complete installation protect the wash basins from any misuse/damage.

## 12. KITCHEN SINK

- Check the rough in dimensions for cold and hot water supply and waste pipe ensuring that all pipe sizing and positions are as per approved shop drawings.

- Identify the model of kitchen sink as per approved shop drawing and carefully transfer the unit to installation location to match with counter top.

- Install the angle valves for hot and cold water supply. Place the kitchen sink in place and adjust its level and location to match with counter top. Support the sink on counter top and apply silicon sealant around sink edges with the counter top.

- Connect the mixer with flexible hose to the angle valve and tighten them with coupling nuts.

- Connect the bottle trap assembly to the sink bowl and fit with the drain pipe in wall and seal the joint with rubber sealing connector of appropriate size.

## QUALITY CONTROL

- QCE along with Project Engineer and site Supervisor will monitor that all components are installed as per the contract specifications and approved submittals.

- Inspection Request (IR) shall be submitted to the Main Contractor/Consultant after completion of Installation of sanitary fixtures.

## SAFETY

- Work will commence as per safety regulations laid down in the contract specification and project safety plan.
- Proper safety harness to be used and secured, if required.
- All personal protective equipment shall be used as appropriate according to the nature of job.
- Housekeeping shall be of good standard and all cut lengths and debris shall be removed.

# Section 12

# Method Statement for Fire Fighting Installation

| | |
|---|---|
| PLANNING DEPARTMENT | |
| Rev.: 0 | **MEP PLANNING MANUAL** |
| Date of Issue: Feb 2007 | |
| Doc No.: MEP-01-07 | |

# Code No. : ME-FF-01

# Method Statement
# Installation of Pipes & Fittings

| | |
|---|---|
| PLANNING DEPARTMENT | |
| Rev.: 0 | |
| Date of Issue: Feb 2007 | **MEP PLANNING MANUAL** |
| Doc No.: MEP-01-07 | |

ARAB MEP
www.arabmep.com

# METHOD STATEMENT FOR INSTALLATION OF PIPES & FITTINGS

## SCOPE AND PURPOSE

This "Method Statement" covers the Installation of Pipes and. Fittings. This will also ensure workmanship and conforms to Contract documents.

## REFERENCE DOCUMENTS

- Project Specifications
- Approved shop drawings (latest revision)
- Approved material submittals

## GENERAL

Installation of Sprinkler and Fire Fighting Pipes and Fittings for the Podium and the Towers, shall be done in accordance with the project specifications and drawings.

## EQUIPMENT

- Electric Pipe Cutter
- Threading Machine
- Grooving machine
- Welding machine
- Grinding Machine
- Chain Block
- Electric Drill and Extension Cable
- Pipe Wrenches & Spanners
- Spirit Level
- Plumb Bob
- Trolley
- Mobile Scaffolding and Aluminum 'A' Ladders

## RESPONSIBLE PERSONNEL

- Project Manager
- Project Engineers
- Site Engineer/Site supervisor
- QA/QC Inspectors
- Safety officer
- Site Foremen

- Pipe Fitter
- Helpers

## METHOD OF PRE INSTALLATION

11. Approved materials shall be purchased in accordance with the approved shop drawings and For Construction Contract Specifications.

12. Proper materials including pipes and fittings are to be selected and used according to approved shop drawing and good engineering practices.

13. Adequate stock of material, tools and tackles, testing equipments and consumables along with sufficient man power shall be arranged to carry out the work.

14. Prior to start of activity, area and access will be inspected to ensure that the area is ready for the work to start.

15. Ensure that all openings are in correct location and as per approved drawings.

16. Ensure that the construction power and water for testing of pipes are provided by the main contactor near to the work place

17. Safe and tidy work place, clear access to the work place, modes to shift material, tools and tackles etc are to be ensured.

18. Gauges used for testing to be calibrated.

## INSTALLATION PROCEDURES

1.  Ensure that all tools needed for Installation are ready.

2.  Layout for Pipe Supports locations based on the approved drawings are by means of a chalk line to ensure straightness & run parallel to alignment of adjacent building surfaces.

3.  Measure locations of Supports and drill concrete slab and fix anchor bolts, continuous thread rod and ring hangers.

4.  Pipe Risers will be supported by means of angle supports anchored to the floor slab or shear wall with u-bolts, nuts and washers.

5.  Cross mains will be supported by means of ring hangers with nuts and washers suspended continuous threaded rod anchored to the concrete ceiling slab.

| | |
|---|---|
| PLANNING DEPARTMENT | |
| Rev.: 0 | |
| Date of Issue: Feb 2007 | **MEP PLANNING MANUAL** |
| Doc No.: MEP-01-07 | |

6. Cross mains will also be supported by rigid supports using steel angles with u-bolts which will be anchored to the concrete ceiling slab.

7. Pipe work will rest freely on supports and aligned properly before final connection.

8. Valves and other in line equipment will be installed where indicated in the approved drawings and as per Manufacturer's recommendation.

9. Install Automatic Air Release Valve at high point of the system as required.

10. All Fire Stopping related works on Fire Protection sleeves are to be done by Main Contractor.

11. Pipe ends and Equipment to be covered with polyethelene cover ensuring that no dirt will go inside.

## JOINTING METHOD:

1. By using Electric Pipe Cutter, cut the desired length of pipe to be joined and ensure the cut end of pipe is square and burr free.

2. For 65mm dia. and larger pipes, cut the pipe to suitable length and roll groove and bevel both ends by grooving machine.

3. For 50mm dia. and smaller, cut the pipe to suitable length and thread both ends by threading machine for fixing threaded fittings.

4. On threaded pipes apply Boss White and Teflon Tape, fix and tighten pipe and fitting using pipe wrenches.

5. Remove all dirt and moisture from pipe ends.

## POST INSTALLATION PROCEDURE

3. Ensure to prepare As-Built Drawings as soon as possible after the work is completed.

## QUALITY CONTROL

• QCE along with Project Engineer and site Supervisor will monitor that all components are installed as per the contract specifications and approved submittals.

• Inspection Request (IR) shall be submitted to the Main Contractor/Consultant during the following stages of work have been completed :-
- Inspection of Piping installation and hydraulic testing

## SAFETY

- Work will commence as per safety regulations laid down in the contract specification and project safety plan.
- Proper safety harness to be used and secured, if required.
- All personal protective equipment shall be used as appropriate according to the nature of job.
- Housekeeping shall be of good standard and all cut lengths and debris shall be removed.

## Code No. : ME-FF-02

# Method Statement
# Installation of Sprinklers

# METHOD STATEMENT FOR INSTALLATION OF SPRINKLERS

## SCOPE AND PURPOSE

This "Method Statement" covers the Installation of Sprinklers. This will also ensure workmanship and conforms to Contract documents.

## REFERENCE DOCUMENTS

- Project Specifications
- Approved shop drawings (latest revision)
- Approved material submittals

## GENERAL

Installation of Sprinklers shall be done in accordance with the project specifications and drawings.

## EQUIPMENT

- Sprinkler Wrench
- Pipe Wrenches & Spanners
- Spirit Level
- Mobile Scaffolding and Aluminum 'A' Ladders

## RESPONSIBLE PERSONNEL

- Project Manager
- Project Engineers
- Site Engineer/Site supervisor
- QA/QC Inspectors
- Safety officer
- Site Foremen
- Pipe Fitter
- Helpers

## METHOD OF PRE INSTALLATION

1. Approved materials shall be purchased in accordance with the approved shop drawings and For Construction Contract Specifications.

2. Proper materials which include all types of Sprinklers are to be selected and used according to approved shop drawing and good engineering practices.

3. Adequate stock of material, tools and tackles, testing equipments and consumables along with sufficient man power shall be arranged to carry out the work.

4. Prior to start of activity, area and access will be inspected to ensure that the area is ready for the work to start.
   ❖ Installed False Ceiling Grids is a pre-requisite for Pendent Sprinkler Droppers to be installed, without Ceiling reference Sprinkler Droppers cannot be installed.
   ❖ Installed Bulk Head Frames is a pre-requisite for Sidewall Sprinklers to be installed.
   ❖ Final Elevation of Air Grilles as per Site Condition is a pre-requisite for Sidewall Sprinklers to be installed.

5. Ensure that all openings are in correct location and as per approved drawings. Properly coordinated Approved Sprinkler Drawings and Approved Reflected Ceiling Plan Drawings is required

6. Ensure that construction power and water for testing of pipes are provided by the Main Contactor near to the work place.

7. Safe and tidy work place, clear access to the work place, modes to shift material, tools and tackles etc. are to be ensured.

## INSTALLATION PROCEDURES

1. Ensure that all tools needed for Installation are ready.

2. Apply necessary Teflon Tapes and Shellac on the Sprinkler thread.

3. For Standard Upright and Pendent Sprinklers, mount the Sprinkler into the 25mm x 15mm reducer finger tight, after ensuring alignment; tighten by using Sprinkler Wrench and Pipe Wrench. Install Sprinkler Guard where required.

4. For Concealed Pendent Sprinklers, mount the Sprinkler into the 25mm x 15mm reducer finger tight, after ensuring alignment; tighten by using Sprinkler Wrench and Pipe Wrench.

5. For Sidewall Sprinklers, mount the Sprinkler (with the inner escutcheon plate in place) into the 25mm x 15mm reducer finger tight, then tighten by using Sprinkler Wrench and Pipe Wrench, ensure alignment by using Spirit Level mounted on the deflector.

   • After Sprinkler is properly installed wrap Sprinkler with masking tape to protect from being painted when Bulkhead and walls are painted.

| PLANNING DEPARTMENT | |
| --- | --- |
| Rev.: 0 | |
| Date of Issue: Feb 2007 | MEP PLANNING MANUAL |
| Doc No.: MEP-01-07 | |

6. Upon False Ceiling, Bulkheads and Walls are finally painted, Concealed Sprinkler Head Covers shall be installed and Sidewall Sprinklers wrapping shall be removed and install Wall Escutcheon plates.

## JOINTING METHOD:

6. Remove all dirt and moisture from pipe ends.
7. Apply Teflon Tapes on Sprinkler thread, fix and tighten and by using Sprinkler Wrench and Pipe Wrench.

## POST INSTALLATION PROCEDURE

4. Ensure to prepare As-Built Drawings as soon as possible after the work is completed.

## QUALITY CONTROL

• QCE along with Project Engineer and site Supervisor will monitor that all components are installed as per the contract specifications and approved submittals.

• Inspection Request (IR) shall be submitted to the Main Contractor/Consultant during the following stages of work have been completed :-
   - Inspection of Piping installation and hydraulic testing

## SAFETY

• Work will commence as per safety regulations laid down in the contract specification and project safety plan.
• Proper safety harness to be used and secured, if required.
• All personal protective equipment shall be used as appropriate according to the nature of job.
• Housekeeping shall be of good standard and all cut lengths and debris shall be removed.

## Code No. : ME-FF-03

# Method Statement
# Installation of Fire Hose Cabinets

| | |
|---|---|
| PLANNING DEPARTMENT | |
| Rev.: 0 | **MEP PLANNING MANUAL** |
| Date of Issue: Feb 2007 | |
| Doc No.: MEP-01-07 | |

# METHOD STATEMENT FOR INSTALLATION OF SPRINKLERS

## SCOPE AND PURPOSE

This "Method Statement" covers the Installation of Fire Hose Cabinets. This will also ensure workmanship and conforms to Contract documents.

## REFERENCE DOCUMENTS

- Project Specifications
- Approved shop drawings (latest revision)
- Approved material submittals

## GENERAL

Installation of Fire Cabinets shall be done in accordance with the project specifications and drawings.

## EQUIPMENT

- Hole Saw
- Electric Drill and Ext. Cable
- Spanners
- Pipe Wrenches & Spanners
- Spirit Level

## RESPONSIBLE PERSONNEL

- Project Manager
- Project Engineers
- Site Engineer/Site supervisor
- QA/QC Inspectors
- Safety officer
- Site Foremen
- Pipe Fitter
- Helpers

## METHOD OF PRE INSTALLATION

1. Approved materials shall be purchased in accordance with the approved shop drawings and For Construction Contract Specifications.

| | |
|---|---|
| PLANNING DEPARTMENT | |
| Rev.: 0 | **MEP PLANNING MANUAL** |
| Date of Issue: Feb 2007 | |
| Doc No.: MEP-01-07 | |

2. Proper materials which include all types of Sprinklers are to be selected and used according to approved shop drawing and good engineering practices.

3. Adequate stock of material, tools and tackles, testing equipments and consumables along with sufficient man power shall be arranged to carry out the work.

4. Prior to start of activity, area and access will be inspected to ensure that the area is ready for the work to start.

5. Ensure that all openings are in correct location and as per approved drawings.

6. Ensure that construction power and water for testing of pipes are provided by the Main Contactor near to the work place.

7. Safe and tidy work place, clear access to the work place, modes to shift material, tools and tackles etc. are to be ensured.

## INSTALLATION PROCEDURES

1. Ensure that all tools needed for Installation are ready.

2. Drill the holes for fixing anchors and bolts. Ensure that the Bottom of the Cabinet is mounted 300mm above the finished floor.

3. Mount the Cabinet, ensuring vertical and horizontal alignments.

4. Mount the Fire Hose Reel into the Cabinet and connect the 25mm dia. Hose to the lock shield valve.

5. Wind the 25mm dia. Hose into the reel

6. Mount the Fire Rack into the Cabinet and connect the 65mm dia. Hose and Nipple to the Pressure Reducing Valve.

7. Hang the 65mm dia. Fire Hose and Nozzle in the Hose Rack properly.

8. Powder Fire Extinguishers beside the 65mm dia Fire Hose in the lower compartment of the Cabinet.

## JOINTING METHOD:

1. Remove all dirt and moisture from pipe ends.

| PLANNING DEPARTMENT | |
| --- | --- |
| Rev.: 0 | |
| Date of Issue: Feb 2007 | **MEP PLANNING MANUAL** |
| Doc No.: MEP-01-07 | |

2.  Apply necessary Boss White and Teflon tape on Threads and install the water supply pipe nipple, pressure reducing valve and lock shield valve by using pipe wrenches.

## POST INSTALLATION PROCEDURE

1.  Ensure to prepare As-Built Drawings as soon as possible after the work is completed.

## QUALITY CONTROL

*   QCE along with Project Engineer and site Supervisor will monitor that all components are installed as per the contract specifications and approved submittals.

*   Inspection Request (IR) shall be submitted to the Main Contractor/Consultant during the following stages of work have been completed :-
    a.  Inspection of Piping Installation
    b.  Witness Hydrostatic Testing
    c.  Final Painting Tuch-up

## SAFETY

*   Work will commence as per safety regulations laid down in the contract specification and project safety plan.
*   Proper safety harness to be used and secured, if required.
*   All personal protective equipment shall be used as appropriate according to the nature of job.
*   Housekeeping shall be of good standard and all cut lengths and debris shall be removed.

## Code No. : ME-FF-04

# Method Statement
# Hydro-Static Testing of Sprinklers

# METHOD STATEMENT FOR HYDRO-STATIC TESTING OF SPRINKLER

## SCOPE AND PURPOSE

This "Method Statement" covers the Hydro-static Testing of Sprinkler & Fire Fighting Piping Network. This will also ensure workmanship and conforms to Contract documents.

## REFERENCE DOCUMENTS

- Project Specifications
- Approved shop drawings (latest revision)
- Approved material submittals

## GENERAL

Hydro-static Testing of Sprinkler & Fire Fighting Piping Network shall be done in accordance with NFPA 13 & 14 requirements & Project Specifications & Drawings.

## EQUIPMENT

- Pressure Pump
- Pressure gauge with Isolating Ball Valve
- Air Release Valve
- 25mm dia. Water Hose
- Aluminum 'A' Ladder & Complete Pipe Fitter's Tools

## RESPONSIBLE PERSONNEL

- Project Manager
- Project Engineers
- Site Engineer/Site supervisor
- QA/QC Inspectors
- Safety officer
- Site Foremen
- Pipe Fitter
- Helpers

## METHOD OF PRE INSTALLATION

1. Ensure that all Sprinkler & Fire Fighting Piping Networks are inspected and accepted by Consultant prior to

| PLANNING DEPARTMENT | |
| --- | --- |
| Rev.: 0 | |
| Date of Issue: Feb 2007 | **MEP PLANNING MANUAL** |
| Doc No.: MEP-01-07 | |

conducting Hydro-static Test.

2. Ensure all Equipment such as Pressure Pump, Tools and tackles, testing equipments and consumables along with sufficient Manpower shall be arranged to carry out the work.

3. Prior to start of activity, area and access will be inspected to ensure that the area is ready for the work to start.

4. Ensure all open ended pipes of the Piping Network to be Pressure tested shall be properly plugged. Install the Pressure Gauge, Air Release Valve and Isolating Ball Valve at the Remotest Point for Pressure reading and Flushing. Connect the Pressure Pump to the System

5. Connect the 25mm dia. Water Hose to the source of Water, with the Isolating Valve at the remotest open, fill the Piping Network with Water and flush the Piping Network until water is satisfactory clear.

6. Shut-off the Isolating Valve, remove the water hose and plug the valve.

7. Safe and tidy work place, clear access to the work place, modes to shift material, tools and tackles etc. are to be ensured.

8. Ensure that Pressure Gauges are calibrated with valid certificates.

## INSTALLATION PROCEDURES

1. Ensure that all Equipment / Tools and Water are ready.

2. Fill up the Pipingl4etwork with Water, then with the use of the Pressure Pump slowly build up the pressure in the System and ensure trapped air is released thru the Air Release Valve. Check for leaks, inspect all Fittings while building-up the pressure.

   Disconnect the Pressure Pump from Piping Network and keep away from Testing Area.

3. Repair leaks and defects, if any, and re-test the Piping Network as follows;
   a. Low Pressure Line at 200 psi (13.8 bar) pressure for 2 hours.
   b. High Pressure Line at 456 psi (31.4 bar) pressure for 2 hours.

4. Ensure that during Hydro-testing, calibrated Pressure Gauges are used as follows;
   a. Range at 300 psi for low pressure line.
   b. Range at 600 psi for high pressure line.

5. Pressure testing documentation shall be submitted along with Inspection request.

6. Fill up the Inspection Request and ensure that all Signatures above Names of Witnesses are taken after satisfactory result is completed.

## QUALITY CONTROL

- QCE along with Project Engineer and site Supervisor will monitor that all components are installed as per the contract specifications and approved submittals.

- Inspection Request (IR) shall be submitted to the Main Contractor/Consultant during the following stages of work have been completed :-
  d. Inspection of Piping Installation and Hydraulic Testing.

## SAFETY

- Work will commence as per safety regulations laid down in the contract specification and project safety plan.
- Proper safety harness to be used and secured, if required.
- All personal protective equipment shall be used as appropriate according to the nature of job.
- Housekeeping shall be of good standard and all cut lengths and debris shall be removed.

## Code No. : ME-FF-05

# Method Statement
# Installation of Fire Alarm System

# METHOD STATEMENT FOR INSTALLATION OF FIRE ALARM SYSTEM

## SCOPE AND PURPOSE

This "Method Statement" covers the site installation of Fire alarm system complete with all relevant devices and accessories and the requirements of checks to be carried out.

## REFERENCE DOCUMENTS

- Project Specifications
- Approved shop drawings (latest revision)
- Approved material submittals

## GENERAL

Fire alarm system integrated with voice alarm system include fire/smoke detection, smoke damper monitoring, emergency voice evacuation system with paging system and firefighter's emergency telephone system. It will be ensure that complete system shall meet the requirements of Local Civil Defence Authority Regulations.

## EQUIPMENT

- Electrician hand tools
- Scaffoldings
- Nylon slings
- Measuring Tape
- Calibrated Megger

## RESPONSIBLE PERSONNEL

- Project Manager
- Project Engineers
- Manufacturer's authorized representative
- Site Engineer/Site supervisor
- QA/QC Inspectors
- Safety officer
- Site Foremen
- Electrician
- Helpers

## METHOD OF PRE INSTALLATION

1. Ensure that approved Materials are available to carry out the work.

2. Fire alarm system components with accessories received at site will be inspected as per approved material submittal. In case of any damage, the same should be brought to the notice of supplier for suitable resolution/replacement.

3. Physical verification of materials will be carried out for any damages prior to taking from stores and also prior to installation.

4. Prior to Commencement of work, areas and access will be inspected to confirm that Site is ready to commence the work.

5. All relevant documentation (Drawings and Materials) and Material applicable to particular section of work will be checked by site Supervisor prior to commencement of work.

6. The site Engineer/ site Supervisor will give necessary instructions to tradesmen (Electricians) and provide necessary approved construction/shop drawings.

7. The site Supervisor/Foremen will also check that proper tools and equipment are available to carry out the work and are in compliance with the contract specification.

8. The site Supervisor will also explain to the tradesmen regarding safety precautions to be observed.

9. The site Supervisor and QC Engineer will ensure that calibrated Megger is available at site for testing.

## INSTALLATION PROCEDURES

1. **General**

   All installations shall be carried out as per the Project specifications and CI/SS/FM the applicable wiring practices as per BS 5839

2. **CABLE INSTALLATION**

2.1 Ensure that cable containment (where applicable) as per approved shop drawing is installed, inspected and cleared for wire pulling.

2.2 Fire alarm cables will be directly cleated to soffit where there is no separate ELV containment is available in the areas covered by false ceiling, in private / core lobbies and car park areas.

2.3 Cable pulling through conduits from the drum end to the other end of the duct manually by using spring wire.

2.4 Ensure sufficient length of cable is maintained to connect to the device as per the approved shop drawings, before cutting the cable on both sides of conduit.

2.5    Upon completion of cable pulling. Inspection Request(IR) will be raised to Main contractor/Consultant.

2.6    Perform insulation resistance test for each segment of cable and ensure the continuity of all cable cores.

## 3.  INSTALLATION OF FIELD DEVICES AND CONTROL PANEL

### 2.1   Manual call points, Smoke detectors, Heat detectors, Strobe light / Sounders:

- Install Manual call points as per approved construction drawings and manufacturer's instructions.
- Terminate cable to the device.
- Check the sound ness of installation and alignment of devices.
- Device addressing by software using auto addressing feature. So separate labeling is not required. This will be carried by Manufacturer's authorized representative.

### 2.2   Monitor Module:

- Install Monitor Module as per approved construction drawings and manufacturer's instructions.
- Check the soundness of installation and alignment of the device.
- All the monitor modules to be installed near by the system from where it gets the input.
- Terminate the cabling to the device.

### 2.3   Line isolator and interface unit for flow switch:

- Install the device as per approved construction drawings and manufacturer's instructions.
- Terminate cable to the device.
- Check the sound ness of installation and alignment of devices.

### 2.4   Fire alarm control panel:

- Check the control panel and its internal components before installation for any damage.
- Install the device as per construction drawings and manufacturer's instructions.
- Check the soundness of installation and alignment of panel.
- Check the battery and its terminals.
- Complete the Cabling terminations (except power supply and battery connection).
- Clean inside the control panel.

## 4.  Power Separation:

- Fire detection cables shall not be placed along side power cables or share the same conduit, channel or sleeve with electrical apparatus.
- Cable runs shall be installed at least 450mm from the nearest source of electromagnetic interference.

| PLANNING DEPARTMENT | |
|---|---|
| Rev.: 0 | |
| Date of Issue: Feb 2007 | **MEP PLANNING MANUAL** |
| Doc No.: MEP-01-07 | |

5. **Labelling :**

- Labels for all cables will be attached according to the specifications. Concealed cabling, due to space restrictions will be tagged with device address at the termination end.

6. **Cable Records:**

- Correct conductor polarity shall be maintained during connection to devices.
- Identification at the Main fire Alarm Panel and associated connector blocks shall be in accordance with standard industrial Practices.
- Prepare as -built drawings after completion of installations to allow Commissioning team to work.

7. **Cable Testing:**

- All cables shall be insulation tested using calibrated instruments.

- Any defects in the cabling system installation shall be replaced in order to ensure complete performance under installed conditions.

- All test results shall be recorded as per the format given and shall be signed by Main contractor and consultant.

## QUALITY CONTROL

- QCE along with Project Engineer and site Supervisor will monitor that all components are installed as per the contract specifications and approved submittals.

- Inspection Request (IR) shall be submitted to the Main Contractor/Consultant during the following stages of work have been completed :-
    - e. After completion of installation of Fire alarm cable for a particular area/section.
    - f. For testing of Fire alarm cables.
    - g. Installation of Field devices particular area or section of works.
    - h. Installation of Fire Alarm Panel.

## SAFETY

- Work will commence as per safety regulations laid down in the contract specification and project safety plan.
- Proper safety harness to be used and secured, if required.
- All personal protective equipment shall be used as appropriate according to the nature of job.
- Housekeeping shall be of good standard and all cut lengths and debris shall be removed.
- Use of proper scaffolding while installing devices on soffit to be ensured.

## Code No. : ME-FF-06

# Method Statement
# Installation of Zone Control Valve Assembly

# METHOD STATEMENT FOR INSTALLATION OF ZONE CONTROL VALVE ASSEMBLY

## SCOPE AND PURPOSE

This "Method Statement" covers the Installation of Zone Control Valve Assembly. This will also ensure workmanship and conforms to Contract documents.

## REFERENCE DOCUMENTS

- Project Specifications
- Approved shop drawings (latest revision)
- Approved material submittals

## GENERAL

Installation of Zone Control Valve Assembly shall be done in accordance with the project specifications and drawings.

## EQUIPMENT

- Chain Block
- Electric Drill
- Hole Saw and Extension Cable
- Pipe Wrenches & Spanners
- Torque Wrench
- Spirit Level
- Plumb Bob
- Trolley
- Mobile Tower Scaffolding & Aluminum "A" Ladders

## RESPONSIBLE PERSONNEL

- Project Manager
- Project Engineers
- Site Engineer/Site supervisor
- QA/QC Inspectors
- Safety officer
- Site Foremen
- Pipe Fitter
- Helpers

| | |
|---|---|
| PLANNING DEPARTMENT | |
| Rev.: 0 | **MEP PLANNING MANUAL** |
| Date of Issue: Feb 2007 | |
| Doc No.: MEP-01-07 | |

## METHOD OF PRE INSTALLATION

1. Approved materials shall be purchased in accordance with the approved shop drawings and For Construction Contract Specifications.

2. Proper materials such as Butterfly Valves, Water Flow Switches, Pressure Gauges, Isolating Gate Valves, Inspectors Test Connections & Drains including pipes and fittings are to be selected and used according to approved shop drawing and good engineering practices.

3. All material, tools and tackles, testing equipments and consumables along with sufficient man power shall be arranged to carry out the work.

4. Prior to start of activity, area and access will be inspected to ensure that the area is ready for the work to start.

5. Ensure that all openings are in correct location and as per approved drawings.

6. Ensure that the construction power and water for testing of pipes are provided by the Main Contactor near to the work place.

7. Safe and tidy work place, clear access to the work place, modes to shift.

## INSTALLATION PROCEDURES

1. Ensure that all tools needed for Installation are ready.

2. Mount the Butterfly Valve on the grooved pipes and fix the couplings and then tighten the bolts and nuts by torque wrench.

3. Measure and mark then drill hole (by hole saw) where the Water Flow Switch is mounted then tighten the bolts & nuts by torque wrench.

4. Before mounting the switch clean inside pipe either side of hole.

5. Measure and mark then drill holes (by hole saw) where the outlet tee for the Inspector Test Connection piping is connected then tighten bolts & nuts by torque wrench. Apply Boss White and Teflon Tapes on the threads and install the Inspector Test Connection & Drain and Union.

6. Measure and mark then drill hole (by hole saw) where the outlet tee for the pipe nipple is connected then tighten bolts & nuts by Torque Wrench. Apply Boss White and Teflon Tapes on the pipe nipple threads and install the Isolating gate Valve and Pressure Gauge.

| PLANNING DEPARTMENT | |
| --- | --- |
| Rev.: 0 | |
| Date of Issue: Feb 2007 | **MEP PLANNING MANUAL** |
| Doc No.: MEP-01-07 | |

## POST INSTALLATION PROCEDURE

1. Ensure to prepare As-Built Drawings as soon as possible after the work is completed.

## QUALITY CONTROL

• QCE along with Project Engineer and site Supervisor will monitor that all components are installed as per the contract specifications and approved submittals.

• Inspection Request (IR) shall be submitted to the Main Contractor/Consultant during the following stages of work have been completed :-
   i.  Inspection of Piping Installation and Hydraulic Testing

## SAFETY

   • Work will commence as per safety regulations laid down in the contract specification and project safety plan.
   • Proper safety harness to be used and secured, if required.
   • All personal protective equipment shall be used as appropriate according to the nature of job.
   • Housekeeping shall be of good standard and all cut lengths and debris shall be removed.

<u>Code No. : ME-FF-07</u>

# <u>Method Statement</u>
# <u>Installation of Pressure Reducing Valve Station</u>

| | |
|---|---|
| PLANNING DEPARTMENT | |
| Rev.: 0 | **MEP PLANNING MANUAL** |
| Date of Issue: Feb 2007 | |
| Doc No.: MEP-01-07 | |

# METHOD STATEMENT FOR INSTALLATION OF PRESSURE REDUCING VALVE STATION

## SCOPE AND PURPOSE

This "Method Statement" covers the Installation of Pressure Reducing Valve Station.. This will also ensure workmanship and conforms to Contract documents.

## REFERENCE DOCUMENTS

- Project Specifications
- Approved shop drawings (latest revision)
- Approved material submittals

## GENERAL

Installation of Pressure Reducing Station shall be done in accordance with the project specifications and drawings.

## EQUIPMENT

- Grinding Machine
- Chain Block
- Electric Drill and Extension Cable
- Pipe Wrenches & Spanners
- Torque Wrench
- Spirit Level
- Plumb Bob
- Trolley
- Grooving Machine

## RESPONSIBLE PERSONNEL

- Project Manager
- Project Engineers
- Site Engineer/Site supervisor
- QA/QC Inspectors
- Safety officer
- Site Foremen
- Pipe Fitter
- Helpers

| | |
|---|---|
| PLANNING DEPARTMENT | |
| Rev.: 0 | **MEP PLANNING MANUAL** |
| Date of Issue: Feb 2007 | |
| Doc No.: MEP-01-07 | |

## METHOD OF PRE INSTALLATION

8. Approved materials shall be purchased in accordance with the approved shop drawings and For Construction Contract Specifications.

9. Proper materials such as PRV's, OS&Y Gate Valves, Pressure Gauges, Isolating gate valves including pipes and fittings are to be selected and used according to approved shop drawing and good engineering practices.

10. All material, tools and tackles, testing equipments and consumables along with sufficient man power shall be arranged to carry out the work.

11. Prior to start of activity, area and access will be inspected to ensure that the area is ready for the work to start.

12. Ensure that all openings are in correct location and as per approved drawings.

13. Ensure that the construction power and water for testing of pipes are provided by the Main Contactor near to the work place.

14. Safe and tidy work place, clear access to the work place, modes to shift.

## INSTALLATION PROCEDURES

9. Ensure that all tools needed for Installation are ready.

10. Mount the Gate Valve on the flange with the gasket and all bolts in place then tighten all nuts by fingers.

11. Mount the pipe Spool with pipe nipple for Pressure Gauge with the gasket and all bolts in place then tighten nuts by fingers.

12. Mount the Pressure Reducing Valve on the flange with the gasket and all bolts in place then tighten nuts by fingers.

13. Mount the Pipe Spool with Pipe nipples for pressure gauge and safety relief valve with gasket and all bolts in place then tighten nuts by fingers.

14. Install the By-pass by repeating items 2 to 5.

15. Tighten all bolts and nuts alternately to ensure alignment by using a torque wrench.

| | |
|---|---|
| PLANNING DEPARTMENT | |
| Rev.: 0 | **MEP PLANNING MANUAL** |
| Date of Issue: Feb 2007 | |
| Doc No.: MEP-01-07 | |

16. Apply Boss White and Teflon Tapes on pipe nipples and install Isolating Valves and Pressure Gauges and Pressure Relief Valve.

17. Support the Pressure Reducing Station properly as specified.

## POST INSTALLATION PROCEDURE

5.  Ensure to prepare As-Built Drawings as soon as possible after the work is completed.

## QUALITY CONTROL

*   QCE along with Project Engineer and site Supervisor will monitor that all components are installed as per the contract specifications and approved submittals.

*   Inspection Request (IR) shall be submitted to the Main Contractor/Consultant during the following stages of work have been completed :-
    a.  Inspection of Piping Installation and Hydraulic Testing

## SAFETY

*   Work will commence as per safety regulations laid down in the contract specification and project safety plan.
*   Proper safety harness to be used and secured, if required.
*   All personal protective equipment shall be used as appropriate according to the nature of job.
*   Housekeeping shall be of good standard and all cut lengths and debris shall be removed.

www.ingramcontent.com/pod-product-compliance
Lightning Source LLC
Chambersburg PA
CBHW030627220526
45463CB00004B/1440